Pluto

Pluto

New Horizons for a Lost Horizon

ASTRONOMY, ASTROLOGY, AND MYTHOLOGY

EDITED BY Richard Grossinger

North Atlantic Books
Berkeley, California

Published by
North Atlantic Books
P.O. Box 12327
Berkeley, California 94712
Cover art by Aphelleon/Shutterstock.com
Cover and book design by Susan Quasha
Printed in the United States of America

"Hades" (pp. 27-32) from *The Dream and the Underworld* by James Hillman. Copyright © 1979 by James Hillman.

Pluto: New Horizons for a Lost Horizon is sponsored and published by the Society for the Study of Native Arts and Sciences (dba North Atlantic Books), an educational nonprofit based in Berkeley, California, that collaborates with partners to develop cross-cultural perspectives, nurture holistic views of art, science, the humanities, and healing, and seed personal and global transformation by publishing work on the relationship of body, spirit, and nature.

North Atlantic Books' publications are available through most bookstores. For further information, visit our website at www.northatlanticbooks.com or call 800-733-3000.

Library of Congress Cataloging-in-Publication Data
Pluto : New Horizons for a lost horizon : astronomy, astrology, mythology / edited by Richard Grossinger.
     pages cm
   Summary: "An anthology about Pluto, encompassing astronomy, mythology, psychology, and astrology with original essays and excerpts of classic required reading about our most famous dwarf planet"—Provided by publisher.
   ISBN 978-1-58394-897-2 (pbk.)—ISBN 978-1-58394-898-9 (ebk.)
   1. Pluto (Dwarf planet) 2. Planets—Exploration. 3. New Horizons (Spacecraft) I. Grossinger, Richard, 1944– editor.
   QB701.P63 2014
   523.49'22—dc23

                                   2014039181

            1 2 3 4 5 6 7 UNITED 18 17 16 15 14
                        Printed on recycled paper

Contents

Introduction

RICHARD GROSSINGER

This anthology puts Pluto into scientific, cultural, and psychospiritual context in conjunction with NASA's *New Horizons* mission, the centerpiece of which is the transit of an Earth-originating satellite with the dwarf planet and its moons on July 14–15, 2015. The dedicated capsule of instruments was launched atop an *Atlas V* rocket at 14:00 EST on January 19, 2006, from Pad 41 at Cape Canaveral Space Center in Florida after an eight-day delay, initially to allow last-minute testing of the rocket's kerosene tank, then to avoid high winds and low cloud ceilings downrange.

It is compelling to picture local weather on Earth affecting the delivery of a package at cloud-less, wind-less Pluto more than nine years later.

A half hour after lift-off, the rocket's Centaur second stage reignited the probe and sent it into a solar-escape speed and trajectory. *New Horizons* crashed lunar orbit in a mere nine hours and was out of the Moon's gravitational field before midnight.

As the scientific package zips 6,200 miles above a bleak Plutonian landscape—the Sun a bright star in perpetual night—Pluto becomes the last and farthest-flung member of the original nine-body Solar System to be imaged close-up through a transiting *Homo-sapiens* lens.

But the book (or digital display) that you hold in your hand is not just a program guide for an engineering feat; it is the pursuit of a mytheme, an irreducible kernel with at least three manifestations: a god, an astrological organizing principle, and a planet. Pluto's mythology, astronomy, astrology, science fiction, and politics form a series of semiotic networks that have to do less with imaging a far-off frigid rock and more with staring into a mirror of our own cosmology in crisis.

Unmanned human assignation with a planet, though a mere mite floats above an orb hundreds of thousands of times its mass and size, is never innocent and without metaphysical consequences. Hours after *Voyager's* close encounter with Uranus's moon Miranda (January 28, 1986), the space shuttle *Challenger* exploded, driving NASA temporarily out of the heavens. Likewise when NASA put *Galileo,* a plutonium-238-driven satellite, on a

108,000-miles-per-hour suicidal plunge into Jupiter's atmosphere (September 21, 2003), the collision sent a black nuclear cloud hurtling down through the planet's cloud layers and changed everything about astrological Jupiter on Earth.

Borderline planet, gadfly, and changeling object, Pluto is Earth's Planet X, a vortex of seeds from beyond space and time, a quixotic 1930s dark body bundling inconclusive physical and paraphysical equations, a mischievous intruder in the zodiac who redefines every chart by the fact that his passage along its threshold evokes the imperceptible shock wave of the system itself.

Pluto also crosses the destiny of the human enterprise at the moment that *its own* crisis transits with Pluto—a synchronicity inside the zodiac, which is Synchronicity Central anyway. Why that played out in interbellum 1930 with precisely that elusive trans-Neptunian pellet, and again at hailing distance in 2015, is a cosmic mudra of an unknown sphinx. But NASA could no more not go after this meme in its last gasp at completing a solar circuit than it could avoid putting a man on the Moon or send robots scuttling across equally inscrutable Martian deserts in their time.

While perhaps little more than an escaped moon or common plutino exoterically, esoterically Pluto is an alchemical *vas hermeticum,* a vessel that, without revealing its nature, translates everything that comes into contact with it into its own icons and symbologies. Plutonian artifacts have a Möbius Strip quality (like a barber's pole of energy) such that their meanings oscillate, in a continuous way, from inside their vessel onto its surface. At Pluto, opposite qualities like joy and despair, utopia and apocalypse, grace and brutality—inside and outside—fuse or exchange properties, as they become momentarily indistinguishable. Much too far away to care, Pluto has bigger fish to fry.

I am reminded of words that transpersonal intelligence Seth invested in our plane via his medium Jane Roberts:

"[T]his dimension [e.g., source realm] nurses your own world, reaching down into your system. These realities are still only those at the edge of the one in which you have your present existence. Far beyond are others, so alien to you that I could not explain them. Yet they are connected with your own life, and they find expression even within the smallest cells of your flesh."[1]

That's Pluto in spades, a deep-lying chimera across which Plutonian sigils project a reality too alien for Earthlings to glean, transmit it right down into their DNA as into the quantum states of their planet's atoms, every particle of dust, grain of sand, and morsel of soil and stone. Even the ordinary Plutonian glob, more than 98 percent nitrogen ice, reinvents itself beyond disclosure of its actual identity or reference point. Science simply doesn't reach that deep into a higher-dimensional universe; it doesn't look far enough into the heavens to see Pluto as more than a pip of old Sol-orbiting rock (though of ambiguous enough caliber to be worth quibbling over). Of course, that's what it is in this dimension or at this astrophysical vibration, but who knows how many dimensions the starry field reflects as it twinkles from an incomprehensibly vaster sky.

For these reasons I have re-scripted NASA's secular Pluto (with its photo op) to go after a more cryptic and numinous object: Body X, bad boy of the Solar System, fluctuating signifier, meta-planetary dweller between planes.

Again Seth, this time Seth II, a higher octave, speaks:

"We do not understand the nature of the reality you are creating, even though the seeds were given to you by us. We respect it and revere it. Do not let the weak sounds of this voice confuse you. The strength behind it would form the world as you know it and sustain it for centuries."[2]

Do not be confused by Pluto's minute size, remote location, pallid signal, and demotion to a dwarf. Honor the mystery instead. It is creating our reality, even this nanosecond.

Meta-Pluto stands for our cardinal node and cosmic key. By its dual existence it expresses our split emanation: self and cosmos. In a realm beyond our understanding, it is generating the zodiac, the starry heavens, and the terms for this world. It is pulling souls into the current universe.

This anthology provides the rubric and premise of that proposition.

I had been tracking the *New Horizons* saga since the mission was announced in 2001, at the time replacing the earlier *Pluto Fast Flyby* as well as its more sophisticated successor, the *Pluto Kuiper Express,* neither of which was ever built or launched. Both of them would have reached Pluto by now, or well before the planet had traveled so far toward aphelion (its farthest point from the Sun) that its gauze-thin atmosphere froze out, i.e., crystallized and dropped to the ground in colorless snow. That process, creeping along an icy, electrostatic lattice day by day, will have wiped any remaining film from the crisp, starry Plutonian night by 2020. (The Earth's atmosphere freezing out, by the way, would mean millennial blizzards depositing miles-high glaciers and icebergs on an airless terrain, then a gigantic white star exploding out of a black sky across a blinding snowpack of dunes covering mountains, valleys, seas, and plains.)

Each mission (*Flyby* and *Express*) was canceled, in turn, for lack of funding, to be replaced by a more efficient concept. Finally, after an intense political battle, NASA got a Pluto line-item approved; hired subcontractors to fashion, program, and forge a metal bird for deployment; and fired the robot into space on a flight plan delivering it pretty much nonstop to where Pluto, traveling in its own eccentric orbit, would cross its path 3,464 days later give or take a fortnight.

NASA's unmanned interplanetary space program took shape in the early sixties with *Mariner's* missions to Mars and Venus. During the heyday of Solar System reconnaissance (from roughly 1979 through a smidgen into the aughts), probes visited Jupiter, Saturn, Uranus, Neptune, and their moons, plus the asteroid Vesta (2011) and various other asteroids and comets. They and hardware from several space agencies—those of the Soviet Union, China, the European Union, Japan, and India—flew by, orbited, and imaged the Moon, and the two planets closer to the Sun than Earth, Mercury and Venus, usually as part of the same mission by using a slingshot effect from Venus to get to Mercury.

The United States's *Pioneer 10* passed within eighty-one thousand miles of Jupiter in 1973, inaugurating a series of ever closer encounters with the region's largest planet—Jupiter is an ideal slingshot for accelerating probes to worlds beyond it too. *Pioneer 11* conducted the first flyby of Saturn in 1979, and two *Voyager* probes soon followed it there. The second, *Voyager 2,* was slung into a pre-programmed trajectory by Saturn's gravitational field so that it intercepted Uranus in 1986 and, after being redirected by Uranian gravity, met Neptune on its ellipse in 1989.

Soon after *Mariner 4* reached Mars, the United States and Soviet Union began parachuting satellites onto Venus (1966). In 1995 the United States sent a probe from the *Galileo* orbiter onto Jupiter where it collected fifty-eight minutes worth of data on exospheric and thermospheric composition, temperature, radiation, band instabilities, and weather while traveling at about thirty miles per second. Before it could sink any further, it succumbed to local ambient pressure exceeding twenty-three Earth atmospheres as well as a meteor-like 307 degrees Fahrenheit.

En route to Jupiter, *Galileo* became the first artificial human object to visit an asteroid (Gaspra) and then the first to detect an asteroidal moon (Dactyl circling Ida).

Ten years later NASA programmed the *Huygens* vehicle to separate from its *Cassini* spacecraft and land in the Xanadu region of Saturn's moon Titan. It settled on a high-albedo elevated plain that had multiple channels running through it.

In November 2014, the European Space Agency separated the *Philae* probe from its *Rosetta* spacecraft and bounced it onto the surface of Comet 67P/Churyumov-Gerasimenko, the first human touchdown on a bolide. Launched in March 2004 from French Guiana, *Philae* preceded *New Horizons* into space by more than a year.

And, of course, humans traveled to and walked on the Moon multiple times during the U.S. *Apollo* program, the first such manned touchdown in 1969.

Only Pluto was missing. It was too out-of-the-way, negligible, and asteroid-like to merit a detour by *Voyager 2* after "Grand Tour" stops at Uranus and Neptune and their moons. On its way to the Kuiper Belt and beyond, *Voyager* didn't even consider Pluto—its Earthbound programmers didn't. Their more exigent priorities were to measure the dwindling influence of the Sun to heliopause, to taste the interstellar medium, and to carry a plaque into Deep Space bearing humankind's message to the universe.

Plus the puny ninth planet wasn't deemed worth the cost of sending a tailor-made payload otherwise.

However, after another decade of scientific advancement, a surprising recognition took hold: the one missing orb was accessible by new, relatively inexpensive technology. Blueprints for Pluto flybys graduated quickly from recreational puzzles to concrete proposals to priority drawing boards. Five years earlier the notion would have seemed both frivolous and implausible. Seven-plus years after a flyby became viable—if still only marginally justifiable on a cost/benefit basis—Pluto had its own select probe en route.

My original query letter for this anthology began as follows:

Dear Colleague,
   Launched in 2006, NASA's *New Horizons* Pluto probe is scheduled to arrive at the planet and its moons in July 2015. This is the last of the original planets in our solar system to be

visited and photographed by NASA satellite. For the record the binary system Pluto/Charon has three other tiny moonlets: Nix, Hydra, and one presently (2012) unnamed but tentatively Cerberus. The barycenter of their orbits does not lie within any single body, so "Pluto" is more properly a swarm. [For the record, the fourth moon is actually Kerberos, and a subsequent fifth was named Styx.]

At North Atlantic Books we are working on an anthology of statements looking ahead this event. Our questions are:

> What do you think the probe will find?
> What would you like it to find?
> What might it find that would change some key paradigm or meaning?

These can be answered on any level that you would like. You can answer one or more of the questions or raise different questions and answer them. You can answer with prose, poetry, or art. You can write on the astrology, mythology, history, or science of Pluto. The final anthology will be a pastiche of different responses and viewpoints that will form an interesting mosaic for viewing both *New Horizons* and the civilization that sent it. The length can be anything from a sentence to ten-plus pages.

We very much respect the scientific viewpoint and will make sure that it is represented with integrity and respect. At the same time we are looking for esoteric, science-fiction, aesthetic, and psychospiritual approaches to the topic.

I received many thoughtful responses ranging from the literal and earnestly scientific (for instance, one-time astronaut Jeffrey Hoffman's "measurements of a thin residual atmosphere" and a College of the Atlantic student's "some never-before-seen quasi-organic self-replicator"), to magical realism and speculative kōans (Jonathan Lethem's "dead pets" and Philip Wohlstetter's "time that Proust lost"), to archetypal induction (Robert Sardello's "interior of matter" and "luminous darkness" and Nathan Schwartz-Salant's "mystery of manifestation"), to science fiction and sci-fi satire (Charley Murphy's "hyperintelligent squid" and "58 genders"), plus medleys of these (Ross Hamilton's "swarming corkscrew/fractal alchemy" and "giant, nearly undetectable transparent quartz sphere owning a small frigid core of rotating parts mimicking Vatican City").

I didn't encourage pets in my call for submissions, but a surprising number of aspirants took that tack—Pluto is a popular tag for recruiting non-humans into civilized society, an ideal blend of consonants and vowels both goofy and frisky, a bit gamy, and bearing a fierce or somber enough legacy to hold its own in the wild. If I had not closed the barn door, a menagerie of dogs, cats, rabbits, birds, and snakes would have run riot through this anthology. My favorite animal submission opened in true Pluto spirit:

> In the spring of 1982 our blond cat "Larch," who not only glowed at night, but also lacking claws, balls, and any farm sense, had lasted less than a month at our new raw land homestead (but, oh was he blissed-out for that brief month, his first away from city streets). We were pretty sure it had been an owl that took him, but it could just as easily have been a red-tailed hawk or golden eagle. Thinking we needed a companion for our black lab, Moonshadow, and a mouser for our rustic tent camp, we quickly found a replacement in the form of a tiny all-black male kitten whom we named "Pluto." [3]

I almost included James Moore's skillfully rendered cats for how they imparted Pluto's dark luminosity as well as its motif of abduction by powerful, paradigm-shifting shapes—in Larch's case a mongrel winged thing from on high. But feline Pluto's otherwise worthy and charming exploits matched this anthology only by synchronicity, or in the spirit that everything is congruent to everything else under, or grazed by, Pluto.

In its realized form, *New Horizons for a Lost Horizon* visits themes and motifs common to Pluto (god, planet, and/or sign), some of them more than others. Though I had considered grouping submissions by subtopics, that would have undermined the reciprocity and synergy of the categories. Discursion about Pluto tends to fuse astronomy, astrology, and mythology, even when its author is not trying to, or trying not to—maybe it's the generic Plutonian weirdness and surrealism.

Central overlapping threads include: (1) the astrophysics, exogeology, and exometeorology of a trans-Neptunian object/dwarf planet; (2) the physics, astrology, and hyper-physics of a complex multi-"planet" swarm in mean-motion resonance; (3) Kuiper Belt landscapes and astronomy; (4) the Pluto/Persephone myth of abduction and commutation as well as a range of psychological parables and complexes underlying and emergent from it; (5) the baseline astrology of Pluto, including its conjunctions on Earth with shadows and traumas, death and resurrection, caves and buried treasures, invisible wealth, and enantiodromia (a principle of equilibrium articulated by Carl Jung whereby a psychological process or cultural event automatically generates its opposite in order to restore cosmic balance); (6) the ritual of assigning the name "Pluto" to the ninth planet; (7) the 2006 redaction of Pluto's status to dwarf planet; (8) realms of hyperdimensionality and hyperobjectivity suggested or prompted by Pluto; (9) surrealism and magical realism around lost objects, exotic creatures, paradoxical emanations, and haunting nostalgias, likewise connoted or triggered by the ninth planet; (10) a "terraformed" stepping stone and gateway linking the Solar System to the outer universe; and (11) a floating orbital chip of ET visitation and planetary engineering, splattered or set in place half a billion years (or even longer) ago.

The latter would have required at least a Type II or perhaps a Type III civilization. On this scale (first proposed in 1964 by the Soviet astronomer Nikolai Kardashev and named after him), Type I civilizations harness the energy output of an entire planet, usually their own, drawing power out of hurricanes and volcanoes while altering tsunamis and earthquakes to their designs and uses. They also likely run fusion engines and photonic star drives.

Type II civilizations harness the energy output of a star, usually their own, and generate about ten billion times the output of a Type I civilization. They probably have perfected antimatter drives and nano-probes. Various *Star Trek* episodes depict a conjectural future Earth as a Type II civilization encountering other Type IIs in their junkets into the universe.

Type III civilizations harness the energy output of a galaxy, or about ten billion times the productivity of a Type II civilization. Having colonized hundreds (or even thousands) of light-years of territory, they draw heat, electromagnetism, fission, fusion, and unimaginable sorts of "burn" from hundreds of billions of stars and dimensional frequencies. To get around their galaxy (or other galaxies), they probably use Planck energy propulsion, quantum entanglement, dark-energy drives, black holes, and matter- as well as information-superposition.

There is no guarantee, of course, that representatives of any of these civilizations even exist, but Pluto and a few other Solar System bodies (some odd-looking asteroids and moons) might be the handiwork or remnants of one of their visitations or regional projects. Cosmic paranoia buffs have it that Churyumov-Gerasimenko is not a comet but an invading or abandoned spacecraft—and hardly a dead or quiet one: the black hulk is dispatching spooky radio signals, beckoning and summoning *Homo sapiens* (or stalking us like a Trojan horse). Else why spend billions of dollars on a 3907-day sally merely to park expensive hardware on a remote random stone? Some conspiracists believe that real danger is involved: "Whatever this object is, it did not ask to be found or scrutinized."

Well, you can look at *Philae's* copious pictures of its "comet," listen to the purported beacon (online), and cast your own ballot: either this is a large, irregular stellar-forged kernel of exogeological popcorn or it is a time-damaged spore of Brobdingnagian interstellar engineering by alien welders, complete with decks, turrets, a wall, and a habitable interior.

Earth, by the way, sits on the Kardashev scale at Type 0, about two hundred years from Type I (if it makes it), a few thousand years from Type II, and about a million years from Type III.

All of this is speculative and extravagant, but exactly the sort of speculation and exotica that Pluto sponsors.

The trans-Neptunian-bound craft was designed and built by Johns Hopkins University's Applied Physics Laboratory. The principal mission investigator was (and is) Dr. Alan Stern, executive director of the Space Science and Engineering Division of Southwest Research Institute in San Antonio, Texas, NASA's subcontractor for the project. Stern and his colleagues conceived a viable and cheap enough program to replace the rejected *Pluto Kuiper Express,* yet slip a similar mission past congressional oversight.

As noted, the *New Horizons* payload was accelerated directly into an Earth-and-solar-escape trajectory so that it didn't get yoked by locally issued gravity to circling its source planet like a telecommunications satellite. Breaking Earth orbit, *New Horizons* eventually attained an Earth-relative velocity of about 36,373 miles per hour (9.9 miles per second), a record for a human-made object.

Here are some snippets worth sharing with Pluto neophytes—experts, be patient; everything below is not common knowledge:

1. The interplanetary satellite doesn't represent a sudden revival of the United States unmanned space program or an *au courant* jaunt to the outer Solar System. *New Horizons* was launched when NASA was at the tail end of its planetary-exploration phase but could still sell (if barely) "frivolous" celestial research to Congress.

   The excursion could *not* have been budgeted today unless a private company like Discovery Channel or Home Box Office undertook it as an entrepreneurial venture, a multi-season outer-space reality show. National commitment and funding are no longer there, in fact even for our country's more utilitarian and downhome infrastructure of roads and bridges.

   The reason that *New Horizons* is so tardy, arriving at Pluto just now, almost ten years after its launch, is a matter of the length of time it takes even an express package to reach the ninth solar orbit from the third. Pluto averages about 3.67 billion miles from the Sun. Because of this immense remove, *New Horizons* is a delayed-delivery event. When it finally arrives, it will seem as though NASA is back in the planetary-exploration game—and space officials and the media will play it that way—but, for the most part, no one will be trying to fool anyone; it's just that genuine explanations based on Pluto's distance, hence *New Horizons*'s late arrival, will be lost or overlooked factoids under the drum-beating and hoopla of a successful mission otherwise (presuming, of course, that the satellite does not encounter any small unforeseeable meteors, nomadic centaurs, or meddling ETs during its last months en route). The public is not attuned to cosmic nuances anyway (real "space-time," orbital geometry, or differential calculus, especially in an immediate-gratification video-game decade), so the reality and pizzazz of Pluto will trump the incidental length and sheer duration of the journey. NASA needs—and should not be begrudged—its additional fifteen minutes, or six months, of fame.

   In another mood Pluto visitation evokes the fifties, as if the craft had been launched from a Flash Gordon spaceport, traipsed unnoticed through heavens of *The Twilight Zone* and *Star Trek,* and emerged unscathed, winging over Plutonian modernity. That dogged malaprop will cling to the grids of raw images from *New Horizons,* as high-pixel post-Hubble astronomy will not be able to scrub off planet Mongo or Queen Fria's ice kingdom of Frigia or (for that matter) traces of Bajoran wormholes or vestiges of the Deep Space Nine outpost once called Terek Nor—but neither will hypothetical Plutonians, Romulans, Sulibans, and Cardassians be able to cancel those relentless nanometers and dot matrices of black dead stone (once they get released from image clean-up).

Oscillations between a geological object and a mythical landscape are intrinsic by now, as they parallel feedback between the astronomical planet and its astrological sign.

2. "Pluto" turns out not to be Pluto. That is, it is not Planet X for which astronomers were searching in February 1930 when Clyde Tombaugh, from a mountain plateau in northern Arizona, found the frozen rock wandering in Gemini and gave it full planetary honors and (later) a Roman name. Thereafter, Pluto and its orbit served as representational proxy, astronomically and astrologically, for the missing Planet X.

3. Though Pluto is technically the outermost "planet," its erratic orbit actually passes inside Neptune's for brief periods; it did so between December 11, 1978 and February 11, 1999, and will next repeat the stunt in 230 years, making Neptune temporarily the outermost original planet again.

4. The name *Pluto* was adopted piecemeal, gradually overtaking more popular alternatives like Lowell (for astronomer Percival Lowell's solution of its orbit), Janus (for its essential dual nature and position facing both the Solar System and whatever lies beyond), Planet X (Pluto's name before it was found, and a nod to its ambiguous status afterwards), and the early favorite, Minerva (for her wisdom underlying Lowell's grail, her companion owl in eternal night, her unfilled feminine slot in the zodiac, and her dramatic cosmological appearance, armed at birth with weapons from the godhead of Jupiter).

However, once "Pluto" stuck, mythological aspects of the Graeco-Roman god of the Underworld poured into the new body. These also imbued mathematical and numerological aspects of the planet's orbit in its conjunctions, oppositions, squares, trines, etc., with those of Sol's other bodies, including the Earth, to produce an overall Plutonian meaning-set and symbology which then interacted with cross-cultural tropes to inspire an esoteric philosophy under the planetary—now minor-planetary—designation "134440 Pluto." A mytheme and esoteric kingdom were born.

All this may have little or nothing to do with the celestial bobbin, which is probably just a large centaur or escaped Neptunian satellite, but—you know something?—they are no longer separable. Even astronomers can't pry them apart.

5. Pluto was *not* named after the Disney cartoon pup. The company dog's name was changed from Rover to Pluto in 1931 to take advantage of the buzz generated by the new planet—corporate America rarely misses a free media bonanza! Ever since, Pluto has been a moderately popular canine and feline name, in part because of the Disney choice and in part because it sounds "cuter" than mere icy space debris or the Lord of Hell.

6. Except for subliminal reasons, it is hard to explain why NASA officials prioritized Pluto for its current "last" unmanned planetary mission given that it is a barren rock with only the remotest chance of life or anything out of the asteroidal ordinary.

At one level, it was probably meticulousness: finishing the house tour of the old Solar System. However, there *are* some regional virtues. Because Pluto lies in the outer rung of the Sun's gyre, far from the hearth, it is a frozen remnant of the System's origin. By contrast, other rocky planets, asteroids, and moons most similar to Pluto in that regard—Mercury, Venus, Earth, Mars, Ceres, Juno, Europa, Callisto, Rhea, Miranda, Triton, etc.—have all been altered to varying degrees by volcanism, flowing liquids, planetesimal and meteor bombardment, habitation, etc.

Meanwhile the gas giants of the outer Solar System are where most of the System's cosmological material not in the Sun itself is stored in gigantic gravitationally-curved deposits: Jupiter, Saturn, Uranus, and Neptune. Pluto more resembles the moons of those worlds than it does the worlds themselves, a factor leading to its demotion.

The astronomical themes associated with Pluto are among those common to the Kuiper Belt more than the Jovian realm. Solid materials and mass thin out as the effects of solar heat and planetary gravity diffuse. The outcome is frigidity, darkness, and geochemical inertia as well as preserved conditions from the original swirl of gas and dust that birthed Sun, planets, moons, comets, meteors, asteroids, planetary rings, centaurs, and assorted anomalous objects. Pluto is relatively unchanged from birth, in suspended animation. It provides proportionately an unaltered, pristine glimpse of an early geological phase of planet formation everywhere.

After all, every planet started out as a molten hydrogen dimple and, before that, galactic dust and, before that, an ember of a star.

So, scrutiny of Pluto is both a late second pass over the satellites of the four gas and ice giants and a first glimpse of Kuiper Belt landscapes and objects (KBOs) and the origin of the System itself. Pluto serves as a prototype, a vestige, a relic, and an archaeological marker of an earlier time.

7. The mission name *New Horizons* was chosen, in part, to match its initials to those of Plutonian moons Nix and Hydra. Decades earlier the name of the planet won out for analogous reasons; its first two letters honored Pickering and Lowell, two astronomers whose quest for Planet X led to Pluto's discovery by Mr. Tombaugh (Lowell preferred the "PL" as "Percival Lowell").

8. The spacecraft carries about an ounce of Tombaugh's ashes, while the dust counter in its payload was named after Venetia Burney, the eleven-year-old English child who won an unofficial sweepstakes by getting a telegraph to Tombaugh (via her grandfather, his colleague) recommending a mythological name that Tombaugh liked enough to sponsor for his pebble.

9. On June 13, 2006, *New Horizons* passed within 63,185 miles of the mile-long asteroid designated 132524 APL by Terrans, and checked it out.

On February 28, 2007, the probe made a near approach to Jupiter, only 1.4 million miles from the System's largest body (next to the Sun) at perijove. The flyby gave it a significant gravitational boost, increasing its speed by about nine thousand miles per hour. The satellite tested its instruments on the Jovian atmosphere, magnetosphere, and moons. Then it went into hibernation, preserving its resources (and appetite) for the main course. Though *New Horizons* was making record time, Jupiter orbits the Sun at a "mere" 484 million miles, so *New Horizons* still had more than three billion miles and eight years of travel to go!

10. *New Horizons* first imaged its primary target, Pluto-Charon, from afar in September 2006. Yet it was not until July 2013 that it drew close

enough to resolve Pluto and its moon Charon into two separate objects. By then radio signals took about four hours to travel from the craft's antenna to stations on Earth.

11. *New Horizons* will spend only about twenty-four hours, one Gaian day, accomplishing its Pluto-related goals as it whizzes past the system. At that point it will be the first Earth mission to a trans-Neptunian world, the first to a double planet, and the first to a Kuiper Belt Object. During *New Horizons*'s roaming peep at Pluto-Charon—sightseeing while conducting forensics at more than thirty-one thousand miles per hour—its paraphernalia will frisk the geology, surface composition, and atmospheric structure of the planet.

   The fact that Pluto even has an atmosphere was first established in the mid-eighties by astronomers observing the dimming rate of stars as they were occulted by the tiny planet passing in front of them. If Pluto had no atmosphere, selected twinkles would have bleared out abruptly. Instead, they went through brief, subtle fades.

   Spectrographic analysis from Earth disclosed a thin envelope of gases—nitrogen, methane, and carbon monoxide—around Pluto, at about one seven hundred-thousandth the atmospheric pressure of Earth. Not much flatus but not entirely waftless either. This exotic haze forms as Pluto approaches the Sun and its ices sublimate while cooling the planet's surface in an anti-greenhouse effect. To an astronaut confronting the likes of negative 382 degrees Fahrenheit, Pluto's ballpark norm, another eighteen degrees either way is probably moot.

As Pluto moves away from the Sun, its gases liquefy and freeze, eventually falling to the surface. As stray quanta of sunlight and charged solar particles find their way there, amplified by the planet's extreme axial tilt and orbital eccentricity, they drive photolysis—chemical conversion—for instance, methane into ethane. Since telescopic observation, this has caused a general reddening of Pluto's surface, brightening of its northern polar region, and darkening of its southern pole. *New Horizons* will also investigate Pluto's largest moon Charon (from pericharon at seventeen thousand miles away) and countless smaller moons (including the presently resolved Nix, Hydra, Kerberos, and Styx). Then it will head toward as-yet-undesignated farther objects in the Kuiper Belt and Oort Cloud: hello-goodbye, Pluto; we hardly knew ye.

Of course, mission parameters and apsides (peripluto, pericharon, perinix, perihydra, etc.) may be altered by NASA as the probe approaches assignation.

You might wonder, with all the resources and fuss invested in *New Horizons*—an epic transit plus advanced strategies for pulling information out of the orbiting object and its suborbital moons—why the spacecraft doesn't just settle into orbit, become part of the regional swarm, and then take its own sweet time harvesting data? Why consume the better part of a decade crossing so many barren leagues to go racing by at metaphorical warp speed?

There is a reason. Given the thrice-postponed launch date, it wasn't possible to store enough fuel on board (using 2006 chemical-propulsion technology) to brake *New Horizons* into a planetary orbit and also get the spacecraft to Pluto before a potential early freeze-out of its atmosphere. In addition, bizarre as it may seem, the slower "approach speed" needed for delicate orbit insertion at a body as small, hence lacking in gravitational heft, as Pluto, would have required almost a decade of additional travel time! *New Horizons* is smoking like a bat out of hell (or the proverbial shit through a goose), i.e., so briskly on a solar-escape trajectory that even a Jovian planet might have trouble snagging it. That is barely fast enough to beat the cosmic clock, yet also too fast to curb at Plutonian scale.

Scale can distort common parameters into astonishing incongruities. How often do fuel weight and critical braking speed add ten years to a journey?

When *New Horizons* was fired into the open universe from Cape Canaveral, Pluto was a full-fledged member of the planetary fraternity, every bit a tantamount comrade in a lodge that included Mercury, Venus, Earth, Mars, Jupiter, etc., but not asteroids or centaurs.

In 2006, soon after the launch, the ninth and outermost world was dictatorially and insultingly "downgraded" from a planet to a dwarf planet (insulting, that is, to NASA, Alan Stern, and Pluto fans everywhere). The demotion undercut a mission already deemed (by some) a scientifically trivial escapade, more a sporting event and nerd master challenge, to put icing on the Grand Tour, than a warranted excursion with reciprocal astronomical payoff—sort of along the lines of, "Why Pluto?" "Because it's there!"

Presto, it's *not* there. A half-pint, hardly worth the cost of the hydrazine, occupies the site of the planet formerly known as Pluto. What an anticlimax! What a comedown!

Since turf wars among scientists are usually waged by proxy or bait-and-switch behind the scenes, we may never know all the motives and nuances triggering Pluto's sudden deplanetization but it was a shot across the bow of Stern-and-crew's probe.

Richard Hoagland considers the denigration an intentional if sly response to *New Horizons*, arising from internecine NASA battles entangled with security-circle cover-ups (as well as a caveat to potential whistleblowers). Readers of his piece can weigh the credibility of such conspiracy and disinformation claims.

In any case, after much debate and with significant dissension, the International Astronomical Union (IAU) finally declared that Pluto could not be a planet, essentially because it was too small and an outlier, though they quibbled for months about how planethood was to be defined so as to account for an existing member's impromptu exclusion and sudden failure to make the cut. Less than 5 percent of the astronomers in the world actually cast ballots, so the verdict was less democratic, at least demographically, than a presidential election decided by the Supreme Court.

If not a subterfuge, then why was Pluto abruptly and peremptorily demoted?

Actually it wasn't abrupt. As Pluto's vital statistics and actual nature were gradually clarified, it seemed less and less like a legitimate member of an exclusive club.

The revaluation began in 1978 when the discovery of Charon allowed the fringe world's indeterminate mass to be calculated at only about one twenty-fifth that of Mercury, the next smallest valid orb. It was less than even Luna, Earth's Moon, and certainly less than other large moons like Jupiter's Ganymede and Saturn's Titan. Its size was, in fact, more in keeping with other distant Kuiper Belt worlds that were being culled from the heavens by increasingly refined astronomical equipment during the nineties and aughts.

More than half a century after Pluto was enrolled, candidates like Quaoar, Varuna, Makemake, and Sedna, were lining up faster than their qualifications could be evaluated. By then a Graeco-Roman Solar System (with Pluto a cherished member) had been in place long enough to consecrate its count, jazz, and mnemonic devices at least in people's minds and loyalties. So, astronomers were paralyzed as to what to do about the gathering throng of Kuiper Belt Objects—trans-Plutonian Pluto surrogates—and their burgeoning controversy.

In addition to its small size, Pluto was suspect because (1) it shares a cluttered orbit with many near twins (dubbed plutinos); (2) it strays from the Solar System's ecliptic, penetrating and leaving it like a sewing needle through fabric; (3) it is really a double-planet with Charon; and (4) as noted, it more resembles the growing group of similar, ever more distant KBOs than its fellow eight full-fledged colleagues.

Plutinos are alternate "Plutos" in the same approximate orbit as Pluto itself, all of them governed by powerful 2:3 mean-motion resonance from Neptune, the nearest orb of any size.

Sixteen plutinos are large enough and bright enough to classify as worlds. Two have planetary names: Ixion and Orcus. Yet like ring fragments around Saturn, the number of smaller plutinos is near limitless.

Then, of the approaching-twenty Pluto-similar bodies that have been found in the farther Kuiper Belt, some are Pluto-size, all are Pluto-like, and most have moons too. No way around it: Pluto belongs with them more than with the gas giants or inner Sol-bathed landscapes.

This sort of planetary reclassification is not unprecedented. The largest asteroid, Ceres, was originally heralded as a planet (1801) insofar as it traveled in the previously empty slot between Mars and Jupiter in keeping with eighteenth-century astronomer Johann Elert Bode's proposed algebraic spacing of the Solar System. Like Pluto, Ceres was spotted when astronomers worldwide were searching for a body in Bode's predicted orbit; that is, for a missing trans-Martian, sub-Jovian world. Yet by 1851 when the number of Cererian "planets" had ballooned to twenty-three with Vesta, Pallas, Hygieia, Juno, and crew, the entire cluster was reclassified individually as asteroids. These planetoids, a better name for them, filled the trans-Martian gap, either the primordia of a stillborn world prevented by King Jupiter from jelling in its vicinity or the debris of a Krypton-like cataclysmic explosion.

For the record Pluto lasted in the planetary lodge (1930 to 2006) twenty-six years longer than Ceres and is also, while we're at it, ten times more massive than the largest asteroid.

Pluto's formal trial began early in 2006 in Prague when the IAU defined a planet as "a celestial body that is in orbit around the Sun, has sufficient mass to assume hydrostatic equilibrium (a nearly round shape), and has cleared the neighborhood around its orbit." Objects fulfilling only the first criterion became known as SSSBs (Small Solar System Bodies); objects (like Pluto) fulfilling only the first two were dubbed dwarf planets.

This dictate was modified on August 16, 2006 to read: "A planet is a celestial body that (a) has sufficient mass for its self-gravity to overcome rigid body forces so that it assumes a hydrostatic equilibrium, and (b) is in orbit around a star, and is neither a star nor a satellite of a planet."

The "improved" definition was quickly revoked, as it would have slid Ceres, Charon, and at least one other Kuiper Belt Object (Eris) into the planetary category—Charon as sharing a double-planet slot with Pluto around a joint barycenter (common center of gravity) rather than in explicit orbit like a satellite. In fact, the planet-moon distinction is relative, for an inevitable shift in the Earth-Luna barycenter will bring it into the definition's double-planet range and make Earth's Moon a planet too (and the Earth a member of a double-planetary system). Even though this will not happen for millennia, the fact that it *will* was considered untenable by astronomers: the Moon is not and cannot be a planet—it would be too weird.

The IAU's main goal at the time was to exclude from consideration planet-near satellites, far Kuiper Belt Objects, bodies smaller than the Solar System's largest satellites, extra-solar planets (soon to be plucked in legion from around other stars), and rogue planetary bodies floating *between* stars.

A final classification of planethood was passed on August 24, 2006, under Resolution 5A of the 26th General Assembly. It read:

> The IAU . . . resolves that planets and other bodies, except satellites, in the Solar System be defined into three distinct categories in the following way:
>
> (1) A planet is a celestial body that (a) is in orbit around the Sun, (b) has sufficient mass for its self-gravity to overcome rigid body forces so that it assumes a hydrostatic equilibrium (nearly round) shape, and (c) has cleared the neighborhood around its orbit. The eight planets are Mercury, Venus, Earth, Mars, Jupiter, Saturn, Uranus, and Neptune.
>
> (2) A "dwarf planet" is a celestial body that (a) is in orbit around the Sun, (b) has sufficient mass for its self-gravity to overcome rigid body forces so that it assumes a hydrostatic equilibrium (nearly round) shape, (c) has not cleared the neighborhood around its orbit, and (d) is not a satellite. An IAU process will be established to assign borderline objects to the dwarf planet or to another category.

(3) All other objects, except satellites, orbiting the Sun shall be referred to collectively as Small Solar System Bodies. These currently include most of the Solar System asteroids, most Trans-Neptunian Objects (TNOs), comets, and other small bodies.

The IAU further resolves: Pluto is a "dwarf planet" by the above definition and is recognized as the prototype of a new category of Trans-Neptunian Objects.

End of the topic for them, but not end of discussion elsewhere.

In truth, this is a meaningless dispute among nomenclaturists, extraneous to the heavens themselves—Pluto knows what it is.

Down here on Earth, it is fair to concede that Pluto may no longer be a planet astronomically, but culturally and historically it *is* a planet—a full-fledged member of the Solar Senate and the last classical, thermodynamically molded orb to be invited into its convocation.

At an early phase I had thought to develop this project in collaboration with NASA. I imagined that the Southwest Institute might actually encourage publicity and general esprit, to celebrate its triumph as well as make *New Horizons* a bit more of an impromptu cultural fête than a managed scientific experiment. After initial enthusiasm, an upbeat phone chat with one of Dr. Stern's assistants, and a few days prior to our scheduled follow-up, I received a terse email: "NASA declines participation in this project."

Whatever the reason, I took it as a combination of no imagination and no tolerance for a funkier universe than their own—no gumption or sense of humor either. But then no one has much of a sense of humor these days— neither fundamentalist scientists nor their fundamentalist foes.

It was naïve of me to expect that space scientists would accept poets, metaphysicians, and astrologers as bedfellows, especially at a time of budgetary scrutiny, political constraints on scientific research, and Tea Party qua Fox Network rubes looking for Benghazi-style gotchas to sling at egghead marks.

Folks within the scientific cabal don't tend to permit cultural imperatives that break with their own ontology and belief systems anyway. Institutionally they have wrenched and pasteurized a laical cosmos out of the incomprehensible complexity, vastness, and majesty of Creation, ignoring (as well) its reflection in Psyche and the deeper realm of All That Is. They have likewise excluded meanings, ethics, values, and consciousness from their authorized reality, consigning them instead to a ghostlike epiphenomenal realm.

Fair enough—that's science's playing field, and no one wants to muddy a glittering hydrogen cosmos with its extraordinary output in the physical plane. But the reason that astrophysicists can't get to the bottom of the whole affair—or through their subatomic particles, quarks, superstrings, amplituhedra, and related tropes to what runs the show from underneath, except for an adventitious Big Bang—is that the reality of the universe is far too multidimensional for their instruments, or even their neurons, to crack.

Most scientists also assume that, despite transient ideological and political hurdles, everything is going their way and in the long run they're on the winning team. Rarely examining the larger picture, the ontological background of their own paradigms and algebraic structures, or the liturgical materialism of the carpet they have rolled over the cosmos, they *ex officio* condone macro-ecological pathologies that their interventions sow across the biosphere. Unexamined technocracy plays a leading role in Earth's present crisis—and that includes rigidly narrow views of space, time, and matter. If life forms are inherently machines subject to terminal dissection, if nature is a resource to be mined and manufactured for ephemeral use, then the threads that swathe the Earth as a living planet begin to unravel.

In the absence of a critical counterview, hyper-rational scientism defines our universe and species plan, which means as well the inability of the experimenter to get his or her ass out of the experiment, let alone track its ripples across subtle biochemical systems, let alone into subtle bodies. All creatures on Earth suffer the imposition of unrepentant materialism on their support systems and critical habitats. This is a primo Pluto/Persephone dilemma.

Of course in their unspoken thoughts scientists know (or intuit) that they are chewing on a mere rind of planets and stars. Their guileless curiosity, wonderment, and awe once drove their career choice and devotion to a deep truth meditation, even if their cabal permits only quotidian recognition of it.

So, despite science's best efforts, astronomy has been inextricably tinged by cultural and aesthetic artifacts. On that basis this anthology is an antidote to sanitized astrophysical memes. Our Pluto is meant to stir far-flung imaginings of not just the late classical planet (or first dwarf planet) but to point symbolically and subconsciously toward hidden and transcendental Plutonian themes, to "new horizons" in alternative technology, cosmology, and ecological and economic awareness. If Pluto can be redefined, so can energy, so can money, so can value and meaning, for mythologically and etymologically the god whose oracle is at "Pluto" happens to be the source and derivation of all wealth, material, and valuation—of materialism itself—as well as the mechanism of any extant trove's reversal and negation. We stand to be taught a lesson by Pluto—by any planet—but Pluto's exercise is particularly paradigm shattering.

I believe finally that the cultural and symbolic effects of planetary exploration are unavoidable. Plus any expression of public enthusiasm for the planets and stars, however misguided it may seem from a scientistic standpoint, in the long run inspires more than stymies celestial inquiry.

In fact, expending resources on Pluto, likely a common Kuiper rock, makes far more astrological than astronomical sense, which may be unconsciously what drove New Horizons in the first place and just as unconsciously persuaded bureaucrats to fund it: you can't underestimate a zodiac armed with synchronicity factors. Pluto the mytheme is a far more active and powerful agent (by comparison to a mere dwarf planet) than realists are able to heel. From a purely astronomical standpoint, it sounds like (and resonates as) a higher roller than it is, in fact to the degree that it fools even scientists and politicians. Pluto walks the halls of NASA and Congress with street cred and glam, just as much of a playuh, diva, and muckety-muck as Uranus, Mercury, and Saturn.

If it were not for these cultural and archetypal attributes on an unconscious level, a rationale for New Horizons would have been hard to construct on a conscious level. If all that was at stake was run-of-the-mill exogeology, why bother? Why valorize Pluto 134340? Why not just go after a nearer asteroid or centaur—less hassle, less travel time, less gas. But meta-Pluto has compiled such an impressive nonscientific resumé that it has taken on its own ineffable pop cachet. That is what makes it seem astronomically more significant than it probably is.

In summarizing the contents of this anthology, I will offer oversight while referencing aspects of contributions (though I will not name every instanced contributor).

Although it makes parochial sense to distinguish astrological and mythological themes from astronomical ones, and to segregate either from vernacular adaptations, these threads (as noted) tend to converge, conflate, and imbricate—unconsciously as well as intentionally, both in the universe and in our texts. Pluto's astrological and mythological characteristics are virtually inseparable from each other and also, remarkably, from Plutonian astrophysics; they inform one another on psychological, symbolic, psychic, and totemic levels.

Though Pluto may be a small, insignificant planet, not even a full planetary body, it is a wide, deep archetype as well as a meta-planetary hyperobject, blending natural and cultural parameters in a tesseract beyond human deconstructability.

Yes, that is another bit of creative hyperbole on my part, but it is the reason that Timothy Morton occupies a key slot in this book. The Plutonian placeholder—the array of parameters and concepts that have emerged from the singular planetoid—have accorded it its special status in the ground of human experience. Morton characterizes the bigger picture as spanning "a geological time (vast, almost unthinkable), juxtaposed . . . with very specific immediate things—1784, 2001, 1943, Hiroshima, Nagasaki, plutonium"[4]— the eponym not irrelevant (plus hyperobjectivity is a fathomless series: global warming, Styrofoam, money, dark matter, torture, McWorld, Jihad, etc.). As cultural rubrics interact with natural objects, both thermodynamically and symbolically, their outcomes spread through space and time in weird, indecipherable ways—storms of multivariate probability waves and their distributions rippling through further transduction fields and their interference patterns.

Morton hypothesizes as well under the epigraph of Percy Shelley's ". . . *awful shadow of some unseen power.*"[5] No matter how many operations, interrogations, digital scans, and deconstructions we perform on Pluto, some aspect of it remains on high: supernal and indecipherable: *"as summer winds that creep from flower to flower; / Like moonbeams that behind some piny mountain shower."*[6] It is as tantalizing as it is ineffable and unattainable.

Hyperobjects are "entities that are massively distributed in time and space, at least relative to human scales. . . . They are viscous, molten, nonlocal, phased, and inter-objective. . . . They appear in the human world as products of our thinking through the ecological crisis we have entered. . . . [T]his is the moment at which massive nonhuman, nonsentient entities make decisive contact with humans, ending various human concepts such as world, horizon, nature, and even environment. Art in the time of hyperobjects isn't simply art *about* hyperobjects but art that seeks to evoke hyperobjectivity in its very form. . . ."[7]

Many of Morton's exemplifications apply to the Plutonian sphere; if not the concrete astronomical body, the combination natural-cultural object and Plutonian archetype:

> One, Viscosity: The more we know about hyperobjects, the more we find that we are glued to them. Hyperobjects adhere to any other object they touch, no matter how hard an object tries to resist. In this way, hyperobjects overrule ironic distance, meaning that the more an object tries to resist a hyperobject, the more glued to the hyperobject it becomes. We find ourselves unable to achieve epistemological escape velocity from their ontological density, just as we were beginning to enjoy our ironic free play. No fair!
>
> Two, Molten Temporality, e.g., Salvador Dali's paintings of melting clocks: Any massive object distorts space-time. Many hyperobjects really are massive enough to do this for real with visible effects as in the case of Planet Earth itself. . . . Hyperobjects are so massive that they refute the idea that space-time is fixed, concrete, and consistent. There is no such thing as a rigid body extended in time and space for this reason. And for every object, there is a radically unknowable space and time because the speed of light sets limits on what objects can apprehend. Hyperobjects end the idea of absolute infinite time and space as neutral containers.
>
> Three, Nonlocality: Hyperobjects are massively distributed in time and space to the extent that their totality cannot be realized in any particular local manifestation. . . . Phenomena such as rain . . . become a local manifestation of nonlocal objects. Thus hyperobjects play a mean trick; they *invert* what is real and what is only appearance. . . . The wet stuff falling on my head is less real than the global warming of which it's a manifestation. . . . Likewise, objects don't feel global warming, but instead experience tornadoes as they cause damage in specific places.
>
> Four, Phasing: Hyperobjects occupy a higher dimensional

phase-space than other entities can normally perceive . . . , which is why they are partly invisible to us 3D humans. They seem to come and go like seasons; yet really they continue to unfold elsewhere than where we look.

Five, Inter-Objectivity: Hyperobjects are shared by numerous entities in a . . . vast nonlocal configuration space . . . of entangling ecological interconnectedness . . . that I call the "mesh. . . ." Hyperobjects are formed by relations between more than one object; consequently, objects are only able to perceive to the imprint, or "footprint," of a hyperobject upon other objects, revealed as information.[8]

It is interesting to note how many of these characteristics apply to scientific and science-fiction speculation about the planet Pluto, to its various astrological and mythological semblances, or to an array of Plutonian proxies. The ninth planet is an object of fascination and controversy because it is intrinsically a hyperobject too—a sinkhole of paradoxes and prophecies.

Pluto's depth and hyperobjectivity jell under the precise terms of the planet's exile from Earth—from terrestrial compasses, biases, and criteria. Because Pluto is far from anywhere, opposite characteristics coexist in a single eerie object. Bring it closer to "somewhere"—anywhere—and its contradictions resolve and show recognizable features accordingly. But Pluto cannot be dragged nearer to Earth by anything short of Type III alien extradition or cosmic catastrophe, so it maintains its improbable nature and remove.

"Lord of the limits and queen of the edge," Pluto wanders along the border between the old Solar System and the new Kuiper-Oort-extended System, as well as along the symbolic (if not thermodynamic) heliopause between the Solar System and the rest of the universe, that is, between the comprehensible and the incomprehensible, the domestic and the alien, between human meanings and the mystery of existence.

At its distance, a *Homo erectus* glyph delivers identical semiology to a *Homo sapiens* cyclotron, and the hoot of an owl is the chant of a monk. From Pluto we can't separate Einstein's equations from Townes van Zandt's lyrics or Kim Kardashian's bare frontal, or any of them from the Bhagavada Gita or the epistles of al-Qaeda and John Keats's "Ode to a Nightingale." You can't tell Wes Anderson from *Vanity Fair* or a trope from a tropism. It's all Earth garbage, Earth static, Earth synapses and logic strings, intraterrestrial output, negative-capability source mode.

Since the 1930s Pluto has been (semi-officially) Earth's "nearest faraway place."

Objectively Pluto presents aboriginal rock, frozen nitrogen, icy climatology approaching the alchemy of absolute zero, abode perhaps of weird life of some sort (in the unlikely instance that its core harbors seismically heated carbon-rich liquids or its surface has been "colonized" from beyond the Solar System). Synchronistically these also reflect Pluto's role as guardian of the Underworld and skipper of the convoy-barge between life and death.

A *"tough-love planetary swarm of moonlets," "dark orphan," "lonely lord . . . / in chthonic darkness / [keeping] his sad, abducted bride,"* Pluto is a dispatcher of *"chilly questions sent out into the galaxy / to travel, perhaps endlessly, / at the speed of light."* (See poets in these pages for the sources of quoted lines).

Pluto long ago emerged from a bottomless cosmic theogony as Robert Kelly's *"Invisible God who ruled / the shadowy land where the dead / in their terrible paradox were / understood to live."* The discovery of Pluto in planetary form awakened its Eleusinian dormancy and tied the orb to an emerging meaning-set in the zodiac, as Pluto took the spot reserved for it by Uranus and Neptune and their Dionysian predecessors.

Pluto's mythological and astrological attributes conjure the bleak side of Shelley's "awful shadow": emptiness, loss, desolation, loneliness, grief, despair, eternal darkness, hopelessness, unconsciousness, cyclonic downdraft, involuntary descent, termless incarceration, loss of ground, trauma, panic, fear, cataclysm, vicissitude, addiction, megalomania, ideological monolith, security state, rape, death of God, loss of soul, leakage of plutonium-238 into the biosphere, warlordism, flash mobs, irregular armies, death terror—the "shadow at the bottom of the world" or, in the words of Robert Kelly, "the darkness that does not precede the dawn but the birth of a radical new order of things."[9]

Yet Pluto's cardinal motifs, equally and conversely, propagate transformation, transcendence, revelation, resurrection, rejuvenation, metamorphosis, heroism, wealth, footing in groundlessness, crucible, intelligible teleology, metadimensionality, numinosity, essence of all beingness, courage, and the basis for moral judgment—the enantiodromic Dreamtime.

In its positive jurisdiction Pluto incubates the pre-Olympian Eleusinian rites as well as timeless antecedents in which Persephone and Demeter perform as operants and agents. These entail, at different tiers, a recovery of the ember of spring from the demise of winter, of the luminosity and situatedness of rebirth from the extinction and amnesia of death—by proxy, the revolt of older, wiser Minoan gods against the tyranny of Zeus and his Olympian entourage: the "eventual integration of anything transcendent."[10] The message of the Eleusian Mysteries and Plutonian latency in Western thought (which goes back well before Persia and Greece, so certainly precedes the discovery of a stray dwarf) is that only by death is resurrection possible; only by submitting to the excruciating gauntlet of dismemberment can one be reborn, can the Soul transcend the body. This transmission of secrets is obscurely petitioned in later ceremonies and initiations like the Rite of Vigil, the Golden Dawn, Woodstock, and Burning Man. It is evoked subliminally and sown throughout the noosphere by *New Horizons*.

When emanating negatively on Earth, Pluto vamps as Hitler, Hiroshima, apocalypse, Mad Max, Boko Haram, the Mexican Zetas, and xenophobia (Pluto the ultimate stranger). Its baleful domain spawns concentration camps, genocides and apartheids, the Hutu *interahamwe,* and ISIS's ravaging pseudo-caliphate—all myrmidons of the Plutonian sphinx: William Butler Yeats's self-arising *"shape with . . . a gaze blank and pitiless as the sun."*[11]

When manifesting positively, Pluto appears as a savior and saint as well as the baseline pivot of another reality, an entirely different universe beyond matter yet accessible to psyche: those are its "infinite directions of inexplicable wonder."

Yet remember the greater paradox, Pluto's Klein-bottle-like topology: absolute positive and absolute negative aspects overlap. Pluto is the European aspect of Shiva, switching robes at the Western phase of the Indo-European continuum. As Shiva/Hades, he is horrific, the most repellent and bloodiest demon of the lot; his blue-throated, snake-garlanded, ash-smeared cynosure, remorseless and impliable, heralds the futility and fate of all material existence, yet somehow extends compassion and hope beyond ordinary purlieus. Shiva is an essential healer and therapist without regard for temporal life and death. S/he carries the message, astonishing when you get right down to what it really says, that death is not death but the opening to an immortality that transcends mortality itself, that dissolves all dialectics and dichotomies.

The psalm that haunts Shelley's lines is Plutonian in origin: that summer winds conceal and cue an older, more serious breeze, a flow between portals of beinghood; that starlight on clouds betokens a universe in which all Souls and creatures are interdependent and all planets and planetoids bear coequal mass.

Pluto's Underworld is finally a birth chamber. Pluto is also a trickster, a clown, a psychopomp like Coyote in indigenous California and the Great Basin—in other cultures, Raven, Mockingbird, Guinea Pig, Fox, Hyena, Badger, Bee, Spider—which means that s/he creates as she destroys, deceives while enlightening, plays pranks and drops misdirects and kōans even as she trains and educates. She dupes shamans as adroitly as she baffles laypersons.

Psyche is always dealing with its own contradictory Plutonian nature, responding to a message as paradoxical as itself. According to astrologer Jeff Green:

> Pluto correlates to the Soul and evolution. . . . [T]he Soul is an immutable consciousness that has its own individuality or identity that remains intact from life to life. In each lifetime, of course, the Soul manifests a personality that has an objective consciousness and unconsciousness. Saturn defines the boundaries of our subjective consciousness—that of which we are consciously aware. Uranus represents the individualized unconscious, Neptune the collective unconscious, and Pluto the Soul itself. . . .
>
> Pluto is a binary planetary system. Pluto's moon, Charon, is actually a planet in its own right. Charon is half the size of Pluto itself, and one twentieth the distance from Pluto as compared to the distance of the Moon from the Earth. The principle of dual coexisting desires seems to be reflected in this planetary symbology.[12]

The fundamental dual nature of Pluto-Charon expresses the original tension between the Soul's desire to remain attached to an impeccable source in a numinous realm—disembodied, enlightened, in a state of eternal epiphany and peace—and its commensurate urge to germinate, incarnate, wake up, gain experience, rough-ride, engage its own true nature. The tension generated by this antithesis explains, even justifies, Pluto's indecencies and damage as well as the most horrific and extreme Plutonian motifs. Pluto is everything and nothing, there and not there, innocent and guilty, latent and manifest, formed and unformable, seraph and abomination, where-the-buck-stops and incapable of responsibility. It changes identities as it approaches the impenetrable basis of its own reality.

Pluto's hubris is *humility itself*.

In this regard it is worth tracking a particularly keen exchange between two noted astrologers, Mark Jones and Steven Forrest, in May 2011. They were riffing off Jeff Green's characterization of Pluto as essential "Soul presence."

Jones introduced the topic by relating the role of Pluto to the lunar south node in a person's chart: "The Pluto level of the past is the more unconscious dynamic to me. I start there because it's an unconscious context, like the deepest security factor. . . . Pluto is where a soulful aspect, a deeper unconscious aspect of a person has been fixated. . . ."

Forrest gracefully accepted the notion of an inner Plutonian fixation but noted that "in [Jeff's] system Pluto represents the Soul." The Soul is not only locked onto Pluto; for all intents and purposes it *is* Pluto—that is, Pluto gathering the moss of negative default patterns from childhood and past lives and general cravings, habits, and other emotional baggage.

While acknowledging that he and Green share a philosophical DNA, Forrest went on to make a key distinction between their reads of Pluto: "To me, the Soul is a mystery." It cannot just be Pluto; it is represented by the entire chart with all its planets, signs, and houses; it also *transcends the chart*. So, he imposed a critical clarification: "Pluto is the *wound* in the Soul." Then he offered Green's comeback: "'Under the patriarchy, all Souls are wounded.'"

Forrest took that as less a rebuttal of his own astrosophical Pluto than a dialectic, a grout melding his and Green's "planets." He put it this way: "Pluto gives us the skills and the powers and the tools that we need to access the wound, and that process is so Soulful and so pertinent to the journey of the Soul that it is equated with the Soul."[13]

Here the matter rests: Pluto is not the Soul, but in this world its action is so deep, so traumatic, so remedial, that its effigy both masks and mirrors the Soul, including the Soul's view of its own situation in us. In other words, we could do worse than to track Pluto as a proxy for the Soul—and astrologers eagerly adopt the matrix.

The core of the Plutonian initiation is that it gets to the bottom of things any way it can, which means that the Soul will always express its core nature.

In the words of another astrologer, Ellias Lonsdale: "Earth is Pluto, and Pluto is Earth. Earth is the depth of Pluto, and Pluto is the mystery of Earth."

Pluto alone dares to break with paradise, with latency—an archetypally perennial act that leads to much subsequent transgression and tragedy. Once cut loose from the Pure Land Buddha-fields with their superheroes, avatars, divinities, exquisite flowers and fruits, ambrosias, and wish-granting trees on which luminous birds glide to rest, etc., the Soul must create an alternate existence in the rough territory of potential and dormant cosmoses. The originary vestal realm surpassed all imagination of beauty but was not resolved or final or *only real*. Its gods were not gods, its flowers not flowers, its birds not birds, at least in the usual sense. Enter Pluto to rake karma and clear matters up.

The Soul must deal with its innate plight, i.e., its curiosity regarding its own nature, views that engender (and embody) whole universes under quantum states and monkey minds, that molt Pluto into its aliases. Lamentably that means conflict, war, atrocities, havoc, mayhem, ingenious instruments of torture, for the Soul's got to know, it's got to know everything about itself and its chattel; that's the only way back, the only reliable course to liberation out of samsara and resolution of harrowing extrinsic pain—either to experience it directly as victim/martyr or to deliver it as perpetrator.

Plutonian duality lies at the heart of the ancient Persephone myth, the coda whereby Jung and his archetypally oriented successors derived an essential transformation: a renewal by metamorphosis in a situation where initially there appears to be only trespass and violation.

While gathering flowers with a sprightly clique from the three thousand Oceanids, Persephone snatched up an especially beautiful and radiant hyper-petalled thing. By uprooting it, she unintentionally opened a cleft, an abyssal chasm to the Plutonian Netherness— in Jungian terms the unconscious foundation of consciousness itself. From there she was forcibly pulled down into a vacuum where she was palpably raped by an invisible god form, a being she could not see. As the force took on quasi-human shape, it compelled her to marry him and relocate to his realm.

Hades (Pluto) had been given prior permission by Zeus (Jupiter) to abduct Persephone because her imperious mother, Demeter (Ceres), goddess of fertility and vegetation, had kept her daughter hidden and unavailable to the gods, most of whom had been smitten by her beauty and charm. Hermes and Apollo, in particular, were beguiled but thwarted from wooing Persephone by Demeter. So, instead, the over-chaperoned girl ended up with the ugly, dark Plutonian shadow.

But Pluto is not just a god or king of the Underworld; he is the astral plane, the fourth dimension, what Rob Brezsny calls "the other *real* world." Pluto is not an abductor but an abyss of pure latency. Likewise in Jungian terms the flower is not just a flower but an opening into her own psyche. That the bloom was a narcissus conflates the myth of Persephone with the archetype of Narcissus, the beautiful son of the river god Cephissus and the nymph Liriope, who became so fixated by the image of his own portrait in a pool that he did not recognize that he was staring at himself (see Nathan Schwartz-Salant's book on narcissistic character disorders, cited in the endnotes).

The loss of her daughter caused Demeter to grieve to such a degree that spring failed to return to the Earth that year, and in ensuing decades. Nothing grew—a perilous state of affairs even for gods insulated on Olympus. After Helios, the Sun, revealed Persephone's location to her mother, Demeter pleaded with Jupiter/Zeus for her release. He responded by allowing the girl, at least in principle, to surface. However, Hades, fully wed and in possession of his prize, was not so easy to juke or rein in. While by the law of the gods he had to obey a command from Zeus, he first enticed Persephone, who was abstaining from food during her ordeal, to swallow several—some say three, others four—pomegranate seeds. Once she had consumed food in the Underworld, she was compelled to dwell there henceforth, at least for part of the calendric cycle, Zeus's sanction notwithstanding—half the year in some versions, a third in others—for the power of the archetype trumps the power of the gods and prevails over our own manipulations, stratagems, and implicit greed too. Persephone's sentence below becomes winter in the world above, a necessary Plutonian cleansing of nature, a rebirth from bared roots and branches to new luxuriant seeds and sprouts.

In older, possibly Minoan versions, Demeter traveled on earth in peasant disguise during her period of mourning and, after helping to raise a royal child incognito, revealed her illuminated and immortal self to the stunned court, then ordered a temple to be built in her name and taught the Eleusinian mysteries there.

At every level, this story is a parable of the psychic forces unleashed by Pluto. The myth is just as cogent in an age when the gods have become stark machineries, diseases, and remote planets that vacated Olympus for the zodiac, e.g., far more obscure but equally sovereign sites from which to rule mankind. The Rape of Persephone is only secondarily and superficially a patriarchal account of plunder and kidnapping. While on the surface an innocent maiden is violated by a self-aggrandizing and antipathetic fiend, that is not the whole story. Pluto/Persephone is a dyad in which each polarity potentiates and makes possible the other. Hades (Pluto) is, according to Rob Brezsny, "a deep, hidden force that manifests throughout the universe on different worlds in different times and is always seminal. You can't have Pluto without Persephone, but you also can't have Persephone unless Pluto awaits her."

In that sense Pluto's power and gift is "a forced initiation which appears to come out of nowhere but changes our lives forever. It is where we are taken over by forces we never believed we were going to get involved with." Pluto's contribution is contravention and rupture; he forces us to go deeper and different. He requires that of us because that's ultimately what and where we are. "If we deny, banish, and repress all the soul's good stuff," Brezsny continues, "the only way it can reach us is through rape and abduction. As its cycle of violence and prurience increases, Pluto imposes a daily punishment, which is not actually a punishment but a necessity, a gift in disguise. The soul will not allow us to forget it, even if it has to express itself by dark acts of requital."

As *New Horizons* approaches Pluto, we see the startlingly sudden appearance on Earth of the planet's shadow and mange, a vast throwback Islamic caliphate, its troops donning the black robes of Mordor and behaving like orcs as explicitly and blithely as if Hollywood had overwritten their script, typecast their actors, and staged their atrocities as a sequel to *Friday the 13th* and *The Texas Chainsaw Massacre*—except that these are scaldingly real deeds from such a naked realm of horror that First Worlders cannot grasp them or their implications, even though Hollywood's own reality show is what is being parodied, hideously.

ISIS is plutonium-238 come of age and incarnate. That is why it shatters its own imagery. The sword drawn anew under Pluto *is the Sword Itself.*

Omar Al-Baghdadi's self-proclaimed Islamic State of Iraq and Syria, or among the infidels, al-Daesh ("the Crushers"), imitates the slasher-and-gore genre despite itself (that is, despite its mock Sharia chastity); then the media picks up on the performance and makes it seem like what is actually happening. That is because the phenomenon has a kernel that is not only imagistically Plutonian but continues to issue subliminal Plutonian commands: life imitating art, art imitating life. ISIS understands that, in a Facebook/Twitter era, the display aspect of their public massacres imposes absolute sound and absolute silence, ripples of which go both forward and backward in time. Fringes dominate the center, and folks find themselves playing characters who have forgotten who they are. It's we who go retrograde, not Pluto. Pluto simply grinds its aboriginal relentless scythe.

For such a ragtag, psychopathic militia, Isis is an oddly unfitting name anyway, except under Pluto. For otherwise, she is the sweetest, most nurturing and obliging goddess in the proto-Egyptian pantheon. *The Daily Spell*, an online magazine of "pagan news and views since 1998" offered this "daily dose of magical thought" in September 2014):

ISIS is not a Terrorist Group in the Middle East
ISIS is the Holy Goddess of Love and Peace
Do NOT use the Goddess' Name in Vain!

ISIS symbolically designates a scorched Earth in global crisis. That it could arise so quickly and unexpectedly, spread so far so fast, and sear such cruelty and sadism so deeply into the human psyche—beheadings, live burials of families, and mass recreational executions—is a reflection of the Plutonian complex and its instantaneity of penetration. Before anyone can react, there it is.

No one even knows the real nationality of ISIS (or ISIL if you substitute "Levant" for "Syria"); it is Iraqi and Sunni but also Chechen, Somalian, Tunisian, Taliban, Pakistani, British, French, German, Canadian, American, Australian. It is a metastasizing street mob of transnational dislocated youth, Islamic jihadists by name but neither Islamic nor involved in jihad; they are really the untagged forces of a disturbed planet seeking a banner and brand under which to coil, unite, and strike back: an aggregate cobra.

It should be no surprise that the United States, Russia, Iran, Israel, and China are equal-opportunity targets of ISIS, hence back-channel allies against it without tacit agreement or openly combining forces.

Introduction

Pluto sets all against itself and itself against all. Its goal is absolute and transmutational; it does what it has to in order to achieve it. It must become ISIS in order to stand against ISIS; it must ravage in order to give birth even as it must adopt the anti-goddess shape of Isis to stand against the godless transnational state. Under Pluto, after all, it is not going to be Tibetan monks leading the way out of pandemic globalism. The serum is going to be a gang straight out of Pluto, live. As Robert Phoenix posts on Facebook:

> Isis is the Mother Of The New Age. Are her progeny Pluto in Scorpio and does that explain the almost demonic savagery they are capable of? In essence, is ISIS more than just terrorizing jihadists? Are they the satanically possessed foot soldiers of Ahriman?

It is not easy for the motley Western alliance to defeat the alliance of an astrological conjunction with an ontological hyperobject, for any attack on it, especially by mere aerial bombing, recoils and releases more of its core symbols, sigils, seeds, and devotees.

Some astrologers propose that as Pluto passes next through Capricorn and repeats aspects of the United States's founding 248 years ago in 1776, it will trigger the beginning of the breakup and coming apart of those same "united states" in 2022–2024—a complete Plutonian cycle. Of course, that is the ultimate goal of the Islamic militants too.

The modern world is full of Persephone-like abductions, from Boko Haram's rounding up whole schools of Nigerian girls and transporting them to a Pluto-like purgatory, to forced marriages of prepubescent children in Pakistan and Afghanistan, to ISIS's raping of captives and selling them into slavery, to the disappearance of girls and young women from the streets of Juarez.

The myth is imposed again and again, sterilely, against the will of Zeus, against the pleas of Demeter, against the Geneva Accords, against the United Nations, against any norms of social justice or internationally accepted taboos. Why? And how do we break the compulsion of negative violative Pluto and enlist positive Pluto's redemption?

The answer lies in Pluto's unexcavated and dual nature.

The planet/god reigns simultaneously over timelessness and time, 248 years plus change per terrestrial-calendar orbit.

For eleven Earth years Cleveland bus-driver Ariel Castro held three kidnapped women captive, much of that time in chains, while sexually molesting them at his whim. Materializing in the camouflage of a suave caring uncle, he abducted them by offering them rides to their destinations, a ruse whereby a modernized Pluto surfaces to woo new victims—"Little Red Riding Hood" the cautionary tale here. We can only guess what flowers distracted them from necessary vigilance, for Madison Avenue and the capitalist desire industry litter Earth's landscape with narcissi.

Nine years prior to being caught, Castro wrote a private memo in which he admitted to being a "sexual predator" and pitifully called out for help. Afterwards he went on molesting those women for approximately 4,730,400 additional minutes. During any of them, he could have stopped on a dime, woken himself up, and pondered, "I have this thoughtform, and it is powerful, and I am acting on it, but it is not who I actually am. I am not imprisoned by it. It is not even my own energy. *Stop, Ariel, stop.*"

But that's how powerful, enduring, and all-encompassing Pluto is. It blots out the stars and sunlight; it turns the universe outside in; it dissolves any foreground in its background mirage. Pluto is not only the compulsion but the myth that binds people to it, that deludes or lulls a person into thinking that it is consummate reality and desire and, without it, he will lose his identity, his core, his gender, and drift in nothingness. It was that imperative that delivered Persephone to Hades in Minoan days, and likely throughout the interminable Stone Ages back to the gods who met humans at the threshold of consciousness itself—that continues to raise a dark mountebank to incarcerate her under its shadow. The fear of the terror of emptiness meets the fear of a prior, *even more terrible* trauma and vacuity.

Castro never progressed through Pluto/Persephone redemption; he held fast to a Charon-the-boatman skeleton in Pluto's frozen realm, forgoing the rites of spring and Eleusinian Mysteries, unaware of the exaltation of release from negative Pluto's grip, the exquisiteness of pleasure that comes with liberation from an ordeal of pleasure-less lust. Then he committed suicide in his prison cell rather than dwell among his own devastating thoughtforms.

Despite appearances, Pluto is actually trying to manifest his opposite: If you *don't* break my compulsion, you will remain in my tidal lock like any orb under gravity, until some other planet comes along and knocks you out.

How drear are Pluto's ongoing fixations?

Mexican cartels specialize in public beheadings too, also live removal of scalps and faces, and castrations, mirrors held up to their bound victims, their own body parts shoved down their throats. Hitmen and enforcers take it as far as their imaginations can—why not, in their handpicked profession of torture and death. But they finally run into the brick wall of materialization itself—that death by Pluto is also resurrection by Pluto, that shamanic extortion is shamanic initiation, that what goes around comes around, karmically. They are holding a mirror to their own faces, in Plutonian time.

A family man and contractor in Tampa, Florida, bamboozled female tourists onto his boat, then one by one raped, tied them up, and threw them overboard alive. A self-denying serial killer on the loose, he was sleuthed down via a public billboard of his handwriting sample (a drama depicted on the television show *Forensic Files*).

How do guys like that think that they can conceal themselves and their actions from God? From Psyche? From Pluto in his dual watch? Do they really think they can hide from the universe and get off scot-free?

In another *Forensic Files* episode a Lubbock rapist strangled his female victims, then calmly slid their bodies into plastic bags and disposed of them in dumpsters. Once again Pluto is in imbalance, negative Pluto turning the screws.

Sad to say, none of these acts, however menial and vile, can arise or operate in isolation, without countless other energies and forms throughout the universe matching them. The zodiac must stay in balance, as big and complex a girth of light and shadow as it is, as patient and wide-ranging an ontology.

Everything Pluto manifests it wrests while standing against itself; everything that Persephone's abduction brings to the Underworld recoils in the Upperworld as something redemptive. When the most depraved crimes and loathsome transgressions hit Pluto, they reverse, convert, and transmogrify. They have to, for there is no place else to go.

Pluto is the wound in the Soul, all right, and these days it is hard to see anything but.

A California man murders his best friend's wife and son in order to kidnap his daughter for whom he has an infatuation. "Where have you gone, James DiMaggio," indeed?

"I haven't gone anywhere," Joe snapped with baffled irritation. He had no terms for recognizing the transformation of the Yankee Clipper into Mr. Coffee.

As for James Lee DiMag, what he accomplished was getting himself turned into a dead star floating in some nameless space, delivering blind and blank energy that continues to need resolution, that is still seeking a stasis and tipping point into something livable. He is at Pluto ultra, Pluto embodied in a Noh play larger than worlds.

The ISIS ravaging of Syria and Iraq, the rampant slaughters of Shiites, Christians, Zoroastrians, Yezidis, and even noncomplying Sunnis; the rampages and kidnappings of Boko Haram; the scourges of the narco-cartels; and Joseph Kony's massacres and kidnappings under his brand of the "Army of God," along with corporate abuses of animals and Earth itself are all Plutonian shadows, recoils through the human psyche of the sociopathic derangement, of soul deadness, of Pluto's toil in the absence of Pluto. Until Pluto arrives to heal it, a Mordor-like plague spreads, birthing dark kingdoms, war crimes, and antipathetic algorithms, for curiosity about our undisclosed nature trumps even the pleasure principle because none of these participants is having real fun.

All their turmoil, baggage, and desperation roils archetypally under NASA's covenant with pure mystery and an essentially undisclosed cosmos— and I don't mean Plutonian body 134340.

It takes more than a space mission and linear exposition to get at Pluto's deepening entry into our midst and influence on the Earth, for we are talking about a planet, a hyperobject, larger than Jupiter, swelling while metastasizing inside both the human and natural sphere. Frederick Ware, an American architect living in France, came close to nailing it in an email to me in lieu of a formal contribution to this anthology. In an exegesis and a dream he foreshadowed Plutonian mayhem:

> Well, Pluto = Hades (minus hermeneutical analyses). Maybe Pluto is a place where one can get some perspective—Out of this World / Nagel's View from Nowhere?
> Had a weird but Big dream this night.
> P O W E R F U L
> (yep. hope it's "only" personal):
> The world's mythical Ocean was in turmoil—i.e., something not seen in memory in that mindspace. A wind of Big Planetary Death swelling to engulf humanity. The oceanic energies were churning, swelling, running over Everything down below. . . .

Travelling with friends—trying to flee—so to find high ground.

I witnessed a Monster wave SWELLING.

It towered above its tumultuous brethren.

Narrative: said to my sidekick, "We gotta get to higher ground!!!" Yeah! (woke up).

Think there's a (not so) hidden meaning with wars, pandemics, ecological disasters just over the horizon? Or just passing stress as silly stupid humanity plods along seeking other refuge for their selfish genes . . . ?

You DID think of the mythic connections with Greek mythopoeic warnings!!??

I respect but doubt primal instincts diffusing from the House of NASA.

Consider that thermodynamical/info-game theoretical energies ANIMATE the spiritual parties in future Olympian temples!

Pluto, as Hades, was master of the Greek endgame. As believer in buddhistic recycling, one sees him as a "business man," collecting alms for passage-coin before taking the boat trip into the bowels of Eternity! All is bardo.

To my lights, key message here—stay @ home, take care of Oikos, home. point!

Better to die extincted—noble or partly race—than survive in future gnostical "worlds," stripped of Dignity, self-serving REPLICONS: Energy slaves. . . .

Replicators—organic Lego bricks/Build the House of Hades. Moral of this cycle's Story?

Just wait for this Mayan cyclic passing . . . cyclic recycler, renewer of Hope.

Believe in US, you, not the Them who pull the strings.

Trying to flee is ULTIMATELY FUTILE!

Serial killers, narcoterrorists, and hooded ISIS soldiers can't be stopped and won't be stopped. Pluto won't have it. Then Pluto *will only have it.*

Pluto is the one planet with enough homeopathic heft, the single world synopsized concisely enough and from far enough away, to convert these and other thoughtforms and release their victims. For the forces that join Ego to Self also pull Ego away from Self, to reveal the rugged reincarnational landscape of the human and cosmic enterprise.

Yes, I know, the rest of us are riding in a presently immune convoy at a safe distance from the outer Solar System and Fourth World, in the upper percentile of the West. But, dudes, here in our privileged digs, we are not immune from Pluto. That is why we presumptuously keep telling the "bad" guys that their Plutos can be transformed into amazing stuff, even in the name of Allah, even while trafficking women and brewing whole drag- ons' lairs of meth, even in the service of fallen angels and avatars, even while mega-guarded in their fortresses. That very Plutonian energy they are hoarding and bowing to, distorting and squandering, holds the key to mercy, divinity, and the actual prana of love.

We rationalize that they are doing the best they can—they are making good use of an impoverished birth in a poor land in a bad time. *We* have the freedom to absolve them as well as our own guilt and vulnerability that dis- pose us to appease them, in our own projections, as if to make a safer, more prosperous world for us all.

We need to go deeper into Pluto in ourselves to scale our own terror and attraction to terror, our apostasy and faux probity against our own dark- ness. Even if the whole world were made of beings like us, even if everyone were like the Dalai Lama, ISIS would still exist, wars would still occur. For everyone has shape-changing Pluto in his or her chart; everyone is matching everyone else's consensus reality in the same zodiac.

Pluto is not only the fact that apocalypse and epiphany are the identical energy at source; it is the precept that you can't get to epiphany without apocalypse and that every act of devastation, every grief, is the crucible of some ambrosial joy. This is a sullen, grim message, but it is the cosmos's one beacon of hope written in the stars: "The Goddess suffers rape, for only in that way does she become transformed by her own mysteries."[14]

Pluto, whoever it finally (with us as its spokesperson), operates beyond its own recognizance and joyless repetition of sterile acts, for it alone has the power to commute them, to deliver us from violence, from betrayal, from gluttony.

It will take hundreds of thousands of years and many lifetimes to get it done Pluto's way, but the universe has at least that much time. That's more or less why it's presently filled with galaxies and stars. The reason that Pluto exists is the reason that *any of this exists.* That was what Seth II was murmuring to our world.

Pluto camouflages the universe from us, but that allows us to be who we are and live our charts without knowing the outcome in advance. Pluto reveals the universe to us at the moment of transfiguration and Soul awakening. That is why the most poignant and delicious emotions we feel are triggered by Pluto's passage: it governs our demise and resurrection, the resolution of our tragedy and karma.

Even Pluto the cat underwent a distinctive Plutonian metamorphosis:

"It was a crisp and cold March morning when she had called our neighbor Ron to bring his .22 rifle and help put Pluto down. They stood in the driveway, with the girls inside perhaps unaware of what was going to happen, and shot the blind old wretched cat.

"The story goes that in a brilliant flash the air became crystal clear, pristine, still, and almost brittle. It was like the limpid clarity behind this world of appearances was revealed. In that moment they both felt an unexpected sense of awe and simplicity and calm. They stood there in that transformed space and hugged each other and cried. In an instant Pluto was gone, and in the release, something remarkable had been revealed. . . .

"It has been many years since these incidences of Pluto's life occurred, yet as I recall them now they bring back strong emotions. Especially remembering the time I almost drowned him, sitting in the car together as we pondered his fate and our connection, it makes my heart open in deep ways, and I find the tears just pouring from my eyes.

"Pluto was more than a cat or companion for those years. He was a glimpse of all of Life. A glimpse of the farther reaches, the extremes of what we all can experience, that are both immediate and tangible, yet also beyond the visible. In the end it seems he was named perfectly." [15]

Named perfectly—the long and the short of it: the *"awful shadow of some unseen power . . . / like memory of music fled. . . ."*

Planet, dwarf planet, criminal, trickster, alien, outlier, banished god, invisible god, deviant lodge member, jihadist, Reiki master, Pluto turns the Solar System into *what it is.*

Pluto's prosperity is intangible and equivocal, bound in the etymology of its name (Πλούτων, *Ploutōn*: giver of wealth), which lies at the root of its god form and authority. "Ploutos" is the most boundless form of affluence because it is generated and allotted value outside sectarian and secular measures as well as beyond the sunny warmth and possessiveness of the ego. It is acquired from the destruction of superficial bounty, transitional opulence and plenitude, whether that be after tragedy and personal death or from the obliteration of entire former star systems into fresh hydrogen gas and dust.

Pluto is made out of, is able to make itself out of, nothing again and again, from time immemorial and *for* time immemorial, in fact beyond the oceans of time. Pluto is the deepest manifest figurement of the Greek pantheon, the guardian of the most profound human and divine wisdom and truth-mystery (in fact, esoteric knowledge is the root of Pluto's older Mycenaean alias, Hades: *Hāidēs*; Doric Ἀΐδας *Aidas*).

The source of not just emeralds, rubies, diamonds, coal, gold, uranium, oil, etc.—things of the Earth—Pluto contains a far more submerged, priceless, and scarce treasure, the matrix of mineral nature itself, the basis of valuation from within planetary cores, under the cosmological ground, back to those molecular clouds of dust and gas, their divine formations and blessings, their subsequent attributes and emanations. Pluto encapsulates the cosmic conversion from gas to liquid to solid and back as well as its paraphysical mutations into plasmas, ectoplasms, and other metaphysical states. These include the miracle of transubstantiating inanimate matter into bacteria, proto-cells, and DNA, then into grains and fruits and creatures. It ultimately governs the terraforming of Pluto itself.

The secret crux of the *New Horizons* probe, the inexpressible, unacknowledgeable motive behind such a Quixotian mission (in terms of our nation's dwindling inflated wealth) is an exploration of our own bare Plutonian nature. The satellite is headed into the human unconscious, into a mirror of us and our global state, into blowback of the scientific hegemony and economic stranglehold on our future and the abandoned richness of our psyche from which new sciences, societies, and truths must someday come. The esoteric side of the *New Horizons* mission is to reveal to us who we are, and it will do so in signature Plutonian fashion, not by explicit image bits, data units, surface measurements, or scant Plutonian facts (engaging though these be) but by seeding its own *fact* in our human and cultural loam, by directly incubating our collective psyche and its transformational archetypes.

Perhaps that is why NASA's subcontractors rejected my invitation to participate in this anthology. They wanted to avoid archetypal dispersal from Pluto itself, subliminally anyway—to remain ignorant of why they were making such a large investment (financial and professional) in such an incidental piece of Kuiper Belt real estate accidentally assigned planetary status because its path crossed that of mysterious Planet X at a time when Clyde Tombaugh was looking into Gemini.

Yet they cannot help being drawn into a Plutonian gyre whereby the hidden planet adumbrates the perceptible one, its shadows, rudiments, and clones inseminating entire realms atop ephemeral placeholders. And that is the case whatever they find there.

This anthology also incorporates science fiction along with scientifically radical appearances of Pluto. Kim Stanley Robinson augurs these themes in his epic trilogy *(Red Mars, Green Mars, Blue Mars)* that concludes with the terraforming of not only our neighboring planet but the entire Solar System, a vision of a future far too akin to Philip K. Dick's dystopias to be as utopian, or ironically utopian, as it sounds:

> The town Hippolyta spanned one of the big groove valleys that were common to all the larger Uranian moons. Because the gravity was even more meager than the light was dim, the town had been designed as a fully three-dimensional space, with railings and glide ropes and flying dumbbell waiters, cliffside balconies and elevators, shoots and ladders, diving boards and trampolines, hanging restaurants and plinth pavilions, all illuminated by white floating lamp globes.[16]

This enticing travel blurb could be taken straight from *Rick Steves's Guide to Pluto and the Outer Solar System.*

Pluto additionally functions as mathematical surd and hidden perturber; as stand-in for the "orbit" of Tyche, a missing suppositional gas giant in the Oort Cloud; as anomalous object; as alien encounter; as interstellar archetype, as dark-matter relic; as black-hole portal; as telekinetic thaumaturge; as ET base; as preserved-ruins and forbidden-archaeology windfall; as lynchpin in the "true history" of the Solar System; as landing platform for ET contact; as orbital message that is more "Pluto code" than "Pluto planet"; as telepathic relay station and interstellar Akashic library for meanings beyond the human operating system or the conventional capacity of bionts.

You'd better picture something out of the ordinary, weirder than crystal domes or celestial factories—not "mere" city-size spaceships, Pandoran dragons, unobtanium deposits, and humanoid robots either but change-lings, shape-changers, poltergeists, cryptids, entities flickering in and out of materiality, rainbow-dimensional portals, qubit nanobots, amplituhedronal quantum-field interactions masking meta-shamanic rituals and nonlocal psychic flying. It's going to be the most extravagant Transdimensional Light and Magic Show ever.

Meanwhile Pluto Coyote prances through rad sci-fi as an artificial body, telltale scrap, and priceless signature of planetary engineering—the fanatical rearranging and remodeling of the size and spacing of the planets and moons orbiting Sol in the original Solar System—by ETs with a very large budget, Type III blueprint, and ambitious utopian plan. Mars's moons Deimos and Phobos have likewise been implicated as artifacts of ET high technology, in fact almost from the instant of their discovery. This alien dossier includes structures in Mars's Cydonia region, craters on Luna, subterfuge asteroids, a chevron shape on the Uranian moon Miranda, and now the Churyumov-Gerasimenko comet.

The most extreme version of this trope is that Pluto is an abandoned relativistic, inertialess-acceleration intergalactic transport machine incorporating the mean-motion-resonance of all its moons and the planetary body itself (merely the largest unit of swarm) into a hyperdimensional drive resembling something out of Douglas Adams's *Hitchhiker's Guide to the Galaxy*.

Finally Pluto is a pellet of the incomprehensible that skirts the human sphere; thus, at every level and interpretation it may not deliver but it always portends, bearing messages and masquerades from the universe at large and from All That Is, from beyond us in our present incarnate form.

How those meanings are voiced by individual authors in this anthology matters little as to details. When you are describing the indescribable, you say whatever you can. For instance, Richard Hoagland's thesis and numerology—Pluto's 248-Earth-year orbit set as an intentional harmonic in a hyperdimensionally resonant Torsion Physics—may not be credible to every reader, and he himself probably wouldn't bet the house on a series of ET "dedicated archives" cached on Mars, Luna, Pluto, etc., or cleverly coded, precisely timed combination information/disinformation modules and a century-long conspiracy and hoax to insert a coincidental bystander in the planetary pantheon as an imposter and Trojan horse. But he plays it as if he would.

In principle and spirit, he is *absolutely* right; he has hit on precisely what Pluto is: the unknown wrapped in the unknowable, a mystery inside an enigma inside another mystery. Any story you give Pluto at the scale of the universe and the distance of any real answer from Earth *fits*.

Notes

Items anthologized in *Pluto: New Horizons for a Lost Horizon* are not noted.

[1.] Jane Roberts, *Adventures in Consciousness: An Introduction to Aspect Psychology* (Needham, Massachusetts: Moment Point Press, 1999; originally published 1975), 5.

[2.] Ibid., 7.

[3.] James Moore, "Pluto," unpublished story, 2012.

[4.] Timothy Morton, *Hyperobjects: Philosophy and Ecology after the End of the World* (Minneapolis: University of Minnesota Press, 2013), 5.

[5.] Percy Bysshe Shelley, "Hymn to Intellectual Beauty," 1816.

[6.] Ibid.

[7.] Timothy Morton, "Dawn of the Hyperobjects," YouTube video, June 16, 2011, www.youtube.com/watch?v=NS8b87jnqnw.

[8.] This is my own conflation of "Dawn of the Hyperobjects" with Timothy Morton's Wikipedia page, http://en.wikipedia.org/wiki/Timothy_Morton.

[9.] A remark made by Robert Kelly in private conversation, ca. 1966.

[10.] Nathan Schwartz-Salant, *Narcissism and Character Transformation: The Psychology of Narcissistic Character Disorders* (Toronto: Inner City Books, 1982), 145.

[11.] William Butler Yeats, "The Second Coming," 1919.

[12.] Jeff Green, *Pluto: The Evolutionary Journey of the Soul,* vol. 1, 2nd rev. ed. (Bournemouth, England: The Wessex Astrologer, Ltd., 2011), 1–2.

[13.] "A Conversation between Astrologers," Forest Astrology News, 2011: http://www.forrestastrology.com/resources/video/video-steven-forrest-and-mark-jones

14. Nathan Schwartz-Salant, op. cit., 151.
15. James Moore, "Pluto," unpublished story, 2012.
16. Kim Stanley Robinson, *Blue Mars* (New York: Harper Voyager, 2009), 532.

1
...

Pluto on the Borderlands

Dana Wilde

The *New Horizons* spacecraft is on a quest for knowledge. It's a robot about the size of a compact car and weighing about a thousand pounds. It uses 228 watts of electricity generated by a radioisotope thermoelectric generator fueled by twenty-four pounds of plutonium-238 oxide pellets and has an array of imaging, radio, and particle and spectrum detection equipment.

Its roughly fifteen-year mission is to fly close to Pluto and its moons and then into the solar system's borderlands, gathering data. It took off from Cape Canaveral, Florida, on January 19, 2006 (the 197th anniversary of Edgar Allan Poe's birth), and about a year later bent around Jupiter where it took some pictures and got a slingshotlike boost that increased its speed to about 52,000 miles an hour. After that it was more or less a straight shot to Pluto, about three billion miles from planet Earth. Its arrival date is July 14, 2015, and the scientists hope it keeps chugging another five years or so into the Kuiper Belt.

It took space scientists nearly two decades to get the mission off the ground because the people who dole out the money for such projects weren't really sure what was to be gained (or who would gain it) by flying to Pluto. It is, after all, far away, and what kind of wealth does an expensive mission like that promise? Even the information you got back would be destined mostly to oblivion in high-powered university computers and esoteric graduate seminars in planetary astronomy. But after a lot of back and forth which included a healthy dose of simple public curiosity, in 2002 the U.S. Congress approved funding for a scaled-down version of the original, early 1990s project proposals.

The mission has three main objectives: to accurately detail the geology and physical structures of Pluto and its moon Charon; to detail and map their surface chemistries; and to detail Pluto's atmosphere. Seems pretty simple. Of course it's not, because so many particles of information are radioing back to Earth that the scientific analysis and conclusions process will take decades. At least.

The physical cosmos is mind-bogglingly detailed. Whole worlds that you can't see are crawling around on your coffee cup rim while you read this, and they all have intricate patterns, structures, chemistries, and interrelationships that took at least three and a half billion years to evolve and have so far taken hundreds of years to track down and describe. The work has barely begun. Even on the next larger scale up from us, meaning the moons and planets, the detail is so fine that just spotting Pluto, which is 1,430 or so miles wide, took decades of hypothesis and observation.

People started speculating seriously in the latter part of the nineteenth century that there might be a planet beyond Uranus (first identified in 1781) and Neptune (tracked down in 1846), and various people, including Percival Lowell of the Martian civilizations theory, invested copious mathematical energy trying to predict where and how large it might be. On the momentum of Lowell's efforts (he died in 1916), the young astronomer Clyde Tombaugh in 1930 found a speck of light changing places from one spot to another on photographic plates, which turned out to be Pluto.

Just a speck of light out there, indistinguishable from millions of other specks unless somebody with a telescope guides you to its exact location. So far away and faint with reflected sunlight that even the Hubble Space Telescope, the greatest device of its kind so far, gets only shadowy images of it.

Curiosity about shadowy, seemingly distant things is always radiating in somebody's mind, including the astronomers who lobbied for the *New Horizons* mission, the mathematicians who wondered if Pluto was there before it was there, and generations of philosophers and mystics whose sense of the cosmic shadows indicated that (for example) "there are innumerable suns, and an infinite number of earths revolve around those suns, just as the seven [planets] we can observe revolve around this sun which is close to us." How Giordano Bruno knew this, or guessed it, decades before Galileo trained a Walmart-quality telescope on Jupiter and its moons, is not clear, but it turned out he was essentially right.

As if the microbes, Earth, and solar system were not detail enough to last an inquisitive species a lifetime, the universe is so vast that there are chips, stones, and grains of dust lying on the surfaces of planets orbiting stars in

mind-boggling numbers. Some astronomers think there cou¹
lion (1019, or 10,000,000,000,000,000,000) planets in the
axy alone. Some guess there could be more than one hundred billᴜ

From this view, Pluto really is a speck of dust. How could the detaiᴌ
geology, chemistry, and atmosphere be so much more important than any othᴄ
grain in the cosmos that a mighty concentration of human mental energy gets
set in motion and crystallizes in a half-ton of carefully configured metals, electricity, and radioactive isotopes and gets sent off specifically to find them out?

Well, one answer is: They're not that important. But another answer is:
They are far away, and shadowy, and reachable.

The scientists put almost all their mental energy into collecting and cat-
egorizing physical facts. But occasionally, when nudged out of their data
dreams, they offer a superficial cliché or two they hope will characterize
the original impulses that drew them there. Barrie Jones, in his book *Pluto:
Sentinel of the Outer Solar System,* states it like this: "Aside from [the] scientific
justification [for sending a robot to Pluto] is the human drive to explore, a
drive that permeates almost all cultures, and has done so throughout human
history." This is a rehearsal of George Mallory's more concise expression of
his reason for climbing Mount Everest: "Because it's there."

By and large the scientists do not say much more than this about it because
their interests, after all, lie in the physical facts, not the metaphysical motiva-
tions. But a conversation has been going on for a hundred years or more—
largely in the academic shadows, as it were—about what effect the scientific
disposition toward what constitutes useful knowledge has on our general
well-being. The hyper-rational philosopher Alfred North Whitehead warned
in the early decades of the twentieth century of the limitations of hyper-
rational thinking. The astronomer Arthur Eddington around the same time
pointed out that, as powerful and practical as scientific inquiry is, there are
nonetheless parts of reality that science cannot treat. Why is a joke funny?
Scientific analysis can't explain it, and annihilates the humor in the effort.

There are other kinds of knowledge besides scientific. We know this.
There is a disposition in science, though, that does not want to talk about
other kinds of knowledge and views them as either beside the point or simply illusory. The nearest many scientists and philosophers of science come
to talking about other kinds of knowledge appears in phrases such as "the
human drive to explore"—we explore because we explore.

Even detailed explanations of what phrases such as "the human drive to explore" mean do not seem well mapped to many scientists. Such explanations depart the world of physical facts too quickly and enter those far-flung parts of reality that science has trouble treating. Giorgio de Santillana and Hertha von Dechend call one of those territories "mythical knowledge." They argue in *Hamlet's Mill* that ancient cultures processed and preserved information about the stars, planets, and their motions through the telling of stories, symbolic rituals, and music. A human being living in such a culture would "have been a participant in the process of mythical knowledge ... In his own person, he would have been part of a genuine theory of cosmology, one he had absorbed by heart, that was responsive to his emotions, and one that could act on his aspirations and dreams. This kind of participation in ultimate things, now extremely difficult for anyone who has not graduated in astrophysics, was then possible to some degree for everyone, and nowhere could it be vulgarized."

In other words, knowledge of the cosmos was not simply a grasp of physical facts and their explanations, but a participation in them through "an idea of the overall texture of the cosmos." That texture is not a rational explanation of physical forces and interactions happening around and to you, but an experience of your place and its activities fitting together with your own mind.

Other scientific-age thinkers who tried to map the experiential territories that seem beyond the borders of scientific inquiry include for example Carl Jung and Joseph Campbell. But despite these forays, we maintain a general cultural assumption that scientific inquiry and analysis is the most highly evolved mental activity human beings can engage in. If prescientific peoples had their own mythologies, those are sort of interesting and quaint and valid *for them*. But *for us* here in 2015, knowledge progresses evolutionarily and rational science is both the pinnacle and the cutting edge of that progression. The mind-boggling beneficences of medicine, communication, and transportation that result from applications of scientific thinking to physical details seem to bear this out. On that momentum, the robot heads to Pluto.

Out there where *New Horizons* is foraging for facts, it is unutterably vast and lonely. Pluto is on the edge of the solar system, meaning that on average it's something over thirty-nine Earth-sun distances (or astronomical units) from the center. Another way of stating its distance and isolation is to observe that, as telescopes got better during the 1980s, '90s, and 2000s, the astronomers grew less sure about what Pluto actually is. It started out as the ninth planet in the 1930s but by 2006 got recategorized—amid considerable dissent—to "dwarf planet" because it turns out that at least one thing out there is bigger than Pluto, and probably more. For the solar system has, in one way of describing it, three parts.

First, there's the solar system of conventional description. This includes the eight major planets (inner planets Mercury, Venus, Earth, and Mars, and outer planets Jupiter, Saturn, Uranus, and Neptune) and their moons. It also includes the asteroids, which are sort of miniature planets and planet chunks orbiting the sun mostly between Mars and Jupiter, but also other areas. In the vicinity of the outer planets they're called centaurs.

Second, just beyond the orbit of Neptune is a borderland called trans-Neptunian space or the Kuiper Belt, or Edgeworth-Kuiper Belt, depending on who's getting credit for predicting its existence. The Kuiper Belt is made up of asteroidlike chunks of rock and ice, collectively called Kuiper Belt Objects, of varying sizes orbiting between about 30 and 50 astronomical units from the Sun. Pluto and slightly wider and more massive Eris are the two largest known KBOs among an estimated seventy thousand with diameters of more than about sixty miles. The most distant large KBO detected so far is a dwarf planet called, as of April 2014, 2012 VP113. Its nearest approach to the Sun is about 80 AU, and the farthest extent of its orbit is estimated to be about 446 AU. It's a little smaller than Sedna, whose closest approach to the Sun is about 76 AU, and at farthest is around 937 AU away. It takes 2012 VP113 about 4,274 years to circle the Sun once, and Sedna more than eleven thousand years.

Third, beyond the Kuiper Belt and Sedna is the Oort Cloud, where an unknown abundance of comets slow-orbit the Sun. No one has ever seen anything in the Oort Cloud because whatever's out there is too small and distant for telescopes—1,000 to 100,000 AU out. But the Oort Cloud is thought to be the origin of long-period comets, meaning the ones that come streaking into our view just once, to return near Earth only after thousands of years, or possibly never.

The farther you get from the Sun, the more frayed everything is. Mercury, Venus, Earth, and Mars occupy orbits relatively compact to the central fire. A giant leap out from Mars, which is on average about one and a half times as far from the Sun as the Earth, the next big thing is Jupiter, looming so huge, there, 5.2 astronomical units from the Sun, that its gravity acts like a shepherd on everything inside the Kuiper Belt. Early on in the formation of our neighborhood, Jupiter's mass (equivalent to about 319 Earths) was so mighty that it pushed Saturn, Uranus, and Neptune into the stable orbits they inhabit now. Saturn, more than three times smaller by mass than Jupiter but still enormous, in turn perturbed Uranus and Neptune, and those two, about seven times and five times smaller by mass than Saturn (but fourteen and seventeen times larger than Earth) swept their neighborhoods clear of many rocky, icy items like or smaller than Pluto—and those are the Kuiper Belt Objects and comets, circling around the Sun like spray on the edge of everything.

After it cruised past Jupiter in February 2007, *New Horizons* went into a hibernation mode with a few wake-up periods to conduct checks. In the next eight years, the empty space it traversed was, for all practical purposes, empty. Uranus and Neptune tooled along in other, far-flung parts of the solar system. The robot is not equipped to detect the scant sort of details lurking in that part of interplanetary space.

About two hundred days before its closest approach to Pluto, *New Horizons* grips down, begins to awaken and starts taking pictures and measurements. In mid-July 2015, it sweeps first within about 6,200 miles of Pluto's surface, and fourteen minutes later within about 16,800 miles of Pluto's large moon, Charon. Radio signals pour data toward Earth. They take more than four hours to traverse three billion miles.

The astronomers were fairly confident about some of the details *New Horizons* would find. Pluto's surface is covered by nitrogen ice, which if you were to walk on it would have a granularity like snow. There are traces of methane and carbon monoxide ices, and just under the surface of craters and otherwise rocky debris is likely to be water ice. The astronomers eagerly anticipate the presence of water anywhere because it implies the presence of processes familiar to Earth. There is a sense, when water is detected or predicted in any quantity, that life is possible—either in the distant past, as on Mars, or even in the present, as underneath Jupiter's moon Europa. No one predicts anything alive on Pluto.

The ices on Pluto's surface vary with its seasons. Like all the planets, Pluto's orbit is elliptical, and also, like Earth in particular, its axis of rotation is tilted with respect to the Sun. These two factors combine to create changing conditions of light and temperature on the surface on both. But on Pluto the effects are extreme. While Earth's tilt is 23.5 degrees, Pluto's tilt is 57.5 degrees, and while Earth's orbital eccentricity (meaning how much the orbit's elliptical shape varies from an exact circle) is 0.0167, Pluto's eccentricity is 0.251. These extreme variabilities, together with the fact that Pluto is so far from the Sun, create distinctly different conditions of light and temperature—different seasons—in Pluto's 249-year year.

Pluto's northern and southern hemispheres tip gradually into and out of the Sun's direct rays just like on Earth, but with steeper shadows at the winter solstice and more direct light on the poles at the summer solstice. The speed of any planet (or dwarf planet) varies according to its location in its elliptical orbit, faster when nearer the Sun and slowing down when farther out. Because of its steep inclination and its variable orbital speeds, Pluto's seasons are not of equal lengths, as they are (roughly) on Earth.

Summer and fall in Pluto's northern hemisphere last about forty-two Earth-years, while winter and spring are about eighty-three Earth-years long. The southern hemisphere, meanwhile, has a shorter spring because Pluto is closer to the Sun during that spell and therefore traveling faster. Because it's so much closer to the Sun during its spring, its temperatures then are probably higher than its temperatures in summer. Its thin atmosphere is thought to expand in summer and shrink in winter in response to warmer, sunnier and colder, darker conditions.

The astronomers treated most of this as fact before *New Horizons*'s encounter.

The scientists are essentially, pragmatically uninterested in how these facts about ice, sunlight, winter, and summer on Pluto fit together with your own mind. But to wander off the beaten way of science for a moment, there is something familiar to someone who lives in, say, Maine about the idea that winter and spring could be nearly twice as long as summer and fall. In the hinterlands of the imagination in the Great White Northeast is the perennial feeling that the stretch of time from November to about early April lasts, well, about eighty-three years, while the stretch from April to October is surely half that, or less. In Maine's deep winter, snow and ice cover everything. The air dries up, the nostrils and throat are perpetually irritated, and human skin turns to ancient papyrus.

On Pluto, similarly, there is more nitrogen snow on the ground in winter because of something similar to dryness, and less on the ground in summer's warmer temperatures because some of the snow sublimates into the atmosphere. Pluto's northern hemisphere will see its winter solstice in the year 2029. Its autumnal equinox was in 1987. On a January afternoon, the beauty at Maine's autumnal equinox seems decades past.

Of course, the average temperature on Pluto's surface is around minus 370 degrees Fahrenheit. Even the bitterest January day in Maine is not quite that cold. Normally. And while in some southern backyard views in Maine the winter sun barely clears the fir tops, at similar latitudes on Pluto it descends more than twice as low because of that 57.5 degree tilt. The shadows there are twice as long and dimmer.

And then there is the matter of Pluto's distance from the Sun. The Earth is 1 astronomical unit—about ninety-three million miles—from the Sun. Pluto is on average about 39.6 astronomical units from the Sun. In Pluto's black sky, with atmosphere too thin to scatter blue light the way Earth's does, the Sun shines eight hundred times less brightly than on Earth. It's a sort of perpetual twilight, somewhat brighter than a full moon, just enough light to read a book. But that landscape of frost and craters, mountains, cracks, and possibly ice geysers would be to an astronaut in some unspoken future invested with a quality of strangeness and remoteness. Like in moonlight, geysers, craters, and frost would appear to lose their actual substance, and become things of intellect alone.

Up above, at least five moons loom. What the smaller ones, Styx, Nix, Hydra, and Kerberos, might look like is uncertain. Bright lights rising and setting. But Charon is another kind of monster. Our Moon spans 0.5 degrees in our sky, large enough to seem mysterious, while Charon spans 3.5 degrees of arc in Pluto's sky because although smaller than our moon, it's much closer to Pluto's surface. And unlike the Earth-moon system, Pluto and Charon are locked in one-to-one orbital resonance, meaning Charon's orbital period is exactly the same as Pluto's rotation period: 6.387 Earth-days. It hangs unmoving over one hemisphere of Pluto.

For in technical terms this is not a planet-moon relationship. It is a double-planet, or maybe double dwarf-planet, system. Charon's diameter is more than half of Pluto's, and it was massive enough at the time its progenitor (most likely) wandered into Pluto's gravitational vicinity and crashed that the center of Charon's orbit is not Pluto itself, but a gravity node just above Pluto, around which they both orbit.

To an astronaut in that unforeseen human future plodding over Pluto's twilight ices, or to the imagination at work in Maine's woods, Charon would be a giant glimmering nightmare. A huge, dimly lit shadow because it reflects so little sunlight, seven times larger in the sky than our moon, looming motionless, caliginous-seeming, waiting. It might become an emblem, even in the mind of an astronaut, of the experience of Pluto's frost-ridden surface, beetling shadows, and icy gleam. Such monstrosities of sky and cold might simply terrify an explorer in a spacesuit.

Or they might not. During that first moon landing in 1969, Buzz Aldrin famously described the lunar landscape as "magnificent desolation," an oxymoron on an experience of profound beauty in perceptually disorienting topography. The astronauts had no context for judging how large or close the lunar boulders and mountains were. The light, shadow, and perspective were strange and remote.

Pluto is much farther away and colder than the moon, much deeper on the edge of science's epiphenomenal inability to connect physical reality to human experience. What a Pluto-walking astronaut might actually experience is not clear. It is not well-mapped territory.

Under Pluto's surface there might be an ocean of liquid water. This hypothetical detail provoked maybe the most excited speculation ahead of *New Horizons*'s encounter. "The subsurface ocean question is central to everything that we don't know yet about Pluto's geology," said one science writer in 2013.

The ocean question is central to everything unknown about Pluto's geology because it would solve the problem of Pluto's internal structure. If inner Pluto is an undifferentiated mix of rock and ices, that's one thing, with one set of implications for how Pluto formed and evolved to its present frozen corpse-like state. But if its interior is differentiated into a water layer and a rocky core, that's something completely different. It means, for one thing, that there is a significant heat source.

Water gets the attention of astronomers and astrobiologists because it is a sort of common denominator. It's central to everything we know about life on Earth. All living organisms, as far as we know, have water at the source of their biochemical processes. We came from water. Our bodies are made mostly of water. One of the clichés that justify the expenses of space exploration is that it helps us learn about ourselves. So water is an elemental link. If an ocean of water exists under Pluto, then we will have found a fragment of our own source reflected there.

The astrobiologists have all but ruled out the possibility of life in any putative Plutonic water. The necessary organic nutrients are thought to have long since leached away. But an ocean inside Pluto could be a sign of water on other large KBOs, like Eris. And they might still have their nutrients. Is there life in the solar system's borderlands?

Key to knowing whether liquid water exists inside Pluto is knowing what process might be melting ice. For there has to be a spark that raises the internal temperature. There are five possible ways planets (and dwarf planets) generate heat: They may have heat left over from the energy released during their crash-battering creation; heat released when denser materials churn downward and less dense materials churn upward; heat released in radioactive decay; heat from tidal distortions like those that the Sun's and Moon's gravities cause the Earth and that Jupiter's causes Europa; and heat released when liquid solidifies.

It is known that Pluto gives off heat. It's measured at a rate about 1 percent that of Neptune. The planetary scientists' prognosis is that the heat is generated either from creation, or from churning materials, or, in particular, from radioactivity. If an isotope of potassium is decaying in the core, then a temperature of around 1,340 degrees Fahrenheit might be generated. Enough heat to keep an ocean (containing some ammonia acting like antifreeze) liquid.

The discovery of a heat source inside a place where hell has otherwise frozen over would constitute a wealth of information.

Pluto's name was suggested by the eleven-year-old granddaughter of an Oxford University librarian. When Venetia Burney Phair's grandfather Falconer Madan read a newspaper account of the new planet's discovery at the breakfast table in March 1930, Venetia, having some knowledge of Greek and Roman mythology as well as some schoolgirl astronomy, said, according to the accounts, "Why not call it Pluto?" Grandfather Falconer's brother had named Mars' two moons, Phobos and Deimos, and liking Venetia's idea, Falconer sent the suggestion up his line of academic contacts. It quickly reached the Lowell Observatory in Arizona, where it had already come up for discussion, and was approved. The ninth planet was Pluto.

Somehow Venetia's guess, or intuition, or epinoia, worked. Pluto is indeed distant and shadowy, qualities we conventionally ascribe to the underworld, whether taken for a real place or a fantasy. The astronomers were pleased with the word *Pluto* because they by and large view the names of astronomical bodies as entertaining whimsies. Scientific common sense presumes that ancient people, bright as they otherwise might have been, were nevertheless vastly ignorant of physical facts and so were restricted to concocting names by fantasy associations. You can find on the Internet, for example, any number of reliable scientific authorities rehearsing the idea that Mars was named for the god of war because it appears red, and red implies blood, and blood implies war.

It's doubtful, however, that the ancient astronomers were really so superficial. An ancient name, like a planet's geology, is nowhere near as simple as its color. Certainly, the ancients were no less intelligent and no less diligent than our scientists, despite our mighty telescopes and advanced mathematics. Cave paintings in southern France that were made thirty-two thousand years ago reveal, at very least, proficiencies of stroke, line, composition, and color to rival any artwork made in historical times. But those ancient geniuses did not possess and were not seeking the same kind of knowledge modern astronomers possess and seek. The astronomers of five thousand and more years ago not only noted a star or planet's color and tracked its path, they also paid attention to the particular kind of awe its luster worked on the mind's eye. They would have been as interested in a Pluto-walking astronaut's apprehensions of looming Charon as the geology beneath his feet. Probably more so.

Venetia, little as she was aware at the time, accurately expressed certain implicit details of Pluto no one else could know either. The name appears "to fit into archetypes that are already there," the astrologer Phil Poirier said in a conversation, and those archetypes involve that which is shadowy, unknown, at the edge of human control, and transitional to previously unseen realities. Was the twentieth century uncovering of horrors, Poirier asked, from the Holocaust to sexual abuse in the church to the burgeoning of mental health therapies, merely coincidental to the discovery of Pluto and its approach to perihelion in 1989? To the scientific mind the answer is resoundingly: Yes, of course. On the other hand, a planet (or dwarf planet) has a presence in the mind beyond its rock and ice. Whether that presence is simply a metaphor, or something more fundamentally interconnected to experience, in either case those mental experiences are more than whimsical associations to a color. They are part of a continuous texture.

"The sky is just the same as the earth, only up above, and older," a Micmac storyteller told an anthropologist around the turn of the last century, distantly echoing a certain kind of knowledge from ancient times.

Pluto does not mean hell, exactly, although that's the conventional notion. It's easy to daydream of metaphorical Charon sculling his boat nearby, waiting. But when you get inside the word, you discover that Pluto is not the underworld place, but instead the ruler himself of Hades. In ancient Greece, the underworld was the place where valuable minerals were found.

"Pluto gives wealth (ploutos)," Socrates explains in Plato's *Cratylus*, "and his name means the giver of wealth, which comes out of the earth beneath. People in general appear to imagine that the term Hades is connected with the invisible, and so they are led by their fears to call the god Pluto instead. . . . In spite of the mistakes which are made about the power of this deity, and the foolish fears which people have of him, such as the fear of always being with him after death, and of the soul denuded of the body going to him, my belief is that all is quite consistent, and that the office and the name of the god really correspond."

To know what's inside Pluto might be central to everything.

You need a telescope to see it. In July 2015, it's nearly 32 astronomical units away and reflects sunlight back at magnitude 14.1. By comparison, Uranus and Neptune, at magnitudes 5.8 and 7.8, are bright enough to be seen on dark, clear nights with binoculars, if you know how to read the star maps.

For us in the early twenty-first century Northern Hemisphere, Pluto is meandering the southern sky in Sagittarius, which is just about the overall most light-filled constellation because in that direction lies the fullness of stars at the galaxy's nucleus. On July 14, 2015, Pluto is a bit northwest of that center. It's just below Venus, which that morning is near its boiling brightest at magnitude minus 4.7. At another time of year you could, with some determination and a four-and-a-half-inch reflecting telescope, spot Pluto near the x Sagittarius stars (36 and 37 Sgr, with naked-eye magnitudes of 5.1 and 3.5). At this time of summer, though, the Sun is about to rise in northern latitudes, and dawn washes out almost everything but Venus.

Even in the telescope accurately aimed from your backyard clearing, Pluto is just a tiny speck of light among a million others. And so no matter whether you can match one crystal-cold dot in a lens to a dot on a star map, or just estimate its general location with your eye, its presence is largely a figment of your imagination. Your imagination can travel in a lot of directions.

The most well worn of those in our eon is mapped by robots. *Voyagers I* and *II* and *Pioneers 10* and *11* years ago cleared the elliptical vicinity of

Pluto's orbit. *Voyager I* in 2015 is beyond the defined range of the Kuiper Belt and drilling along through interstellar space. When your mind is imagining it out there like this, you can kind of keep a grip simultaneously on the feelings of awe and trepidation that thoughts of deep space inspire and on the basic physical data, such as the figures that describe distances: *Voyager I* in March 2014 was more than 127 astronomical units from the Sun. That is a graspable number reflecting a chilling kind of isolation.

At some indefinite psychological intersection, the objective fact meets the mind and gets a shape. One of the shapes is rational: It goes into a category and becomes part of a description of a physical process or condition. Another of the shapes is emotional: the image of deep space inspires a feeling of cold awe, possibly fear.

Science eschews emotions. They are obstacles to rational clarity. Trepidation, awe, fear, hope, desire, pity, animosity, avarice and so on act like smears on the telescope lens of rational thought and you can get lost in them as if in pathless woods. You do not want your astronauts walking around on the moon, let alone on Pluto, frightened out of their wits.

On the other hand, part of the reason you go out to the backyard with your little reflecting telescope is to put yourself in the way of awe and trepidation. You can't, after all, gather any details that the astronomers with powerful robots haven't already got.

The Pluto of awe and trepidation is a shape in the mind that might have gripped Edgar Allan Poe's imagination. Floating up there somewhere in Sagittarius, a figure of darkness, shadows, and icy desolation. For Poe, beyond the fundamental, even adolescent-seeming horrors of his stories was one of the Western world's important proto-psychologists. He made some of the first modern maps of differentiations in the human mind, predecessors of Freud's conscious/unconscious layers and Jung's core archetypes and collective unconscious. Poe's maps of our mental geology show that the experience of Truth is processed through the Intellect, whose faculty is Reason; the experience of Duty is processed through the Moral Sense, whose faculty is Conscience; and Beauty is processed through the Soul, whose faculty is Taste; what Poe calls Passion is parallel to what he, and we, somewhat nebulously refer to as the Heart.

In Poe's view of the human mind, the objective, scientific facts of Pluto might inspire feelings that correspond, in a phrase made by an explorer, to magnificent desolation, where the intellect grasps the desolation, and the faculty of taste (in its nineteenth century sense) experiences the magnificence.

Astronomers, however, bypass these experiences as besmearing emotions, and neuroscientists discard this kind of psychology as, well, immaterial. For the only concrete evidence of mental or emotional experience is in tabulations of biochemical activity. And since the only physical evidence of the existence of mental activity is in chemicals, it is confidently assumed that the chemicals generate the activity. In other words, thoughts, emotions, and awareness itself are produced by biochemical activities. Consciousness is a byproduct of, or epiphenomenal to, chemical processes.

But if chemical activity in the brain is generating the awareness of physical perceptions such as cold, starlight, salt, thunder, and sulfur, then what is happening when the objective facts meet experiences and faculties inhabiting reaches of the mind beyond the range of raw emotion? If outer physical events trigger chemical reactions that result in sensory perception, then what triggers the chemical reactions that register nonsensory, mental experiences such as beauty?

Plenty of scientists, going back at least to Whitehead and Eddington, know very well that the five senses detect nowhere near the whole range of physical reality. We perceive only a tiny fragment of all the wavelengths of light that comb the universe; no one has ever smelled a quark. Despite the best efforts of modern science to deny or ignore them, signals other than five-sensory are coming out of the cosmos, and religious adepts explored and elucidated their sources long before Poe.

The astronomer Robert Jastrow a few decades ago gave a parable of scientific knowledge's link to religious knowledge: A modern scientist after a lifetime of work finds he "has scaled the mountains of ignorance; he is about to conquer the highest peak; [then] as he pulls himself over the final rock, he is greeted by a band of theologians who have been sitting there for centuries."

Once you climb out of the foothills, names are no longer whims, and metaphors are no longer toys. About two-thirds of the way up Mount Purgatory, Dante turns to speak to his guide, Virgil—who signifies the strength of rational mind required to navigate the horrors of hell and the disciplines of purgation—only to find that the guide has peeled off and disappeared because his rational intelligence serves no purpose farther up. There is another kind of knowledge at the mountaintop, and it is encountered in Beatrice.

Beyond the emotional and then beyond the rational are ranges of experience in a psychic borderland difficult to describe because it is, in a way, so distant from everyday perceptual reality. It is for the most part so far off scientific maps that neuroscience denies it even exists. Its images seem shadowy and disorienting. It can be described here no more clearly than as the silence that emerges when noise subsides, or as the realization that the image in a mirror, very clear to your physical eyes, is empty. Not the rational or figurative concept that the mirror is empty, but the realization of its emptiness.

Out of the sky and down through the firs into the tube of the little telescope flows starlight. It strikes a mirror at the bottom of the yard-long cylinder and bounces back up onto another mirror that reflects it again into a lens that funnels photons into your eye. Inside your skull the light signals transpire into a texture of the cosmos. It's a state of mind that stargazers deliberately seek.

At first, on a July pre-dawn, that texture is experienced like a door opening into fresh air. It unfolds into the feeling, maybe, that what's above is the same as what's below, only older. Much older. The overarching feeling, whatever else goes with it, is that the sky is unspeakably beautiful.

The light you're seeking this July morning is hidden in the borderlands of the solar system among a millionfold fullness of vastly more distant lights. Sun rays travel three billion miles, strike Pluto's frost surface and bounce back Earth-ward to a four-and-a-half-inch mirror pointed toward the galactic center. A tiny point of reflected light pierces your eye.

Streaking past Pluto this morning is a robot so tiny in comparison to a moon that its reflection can't be detected by telescope. But built into it is an artificial life of radioactivity and electricity configured to collect signals from the dwarf-planet system and send them on beams of light back to Earth with arcane mathematical coherency. It's the same coherency that emanates us-ward in starlight, arranged so it corresponds to the wavelengths we detected first by eye and then deduced are signs of further wavelengths that transcend the eyeball. It's as if we sent a replica of ourselves out there. The robot picks up data and reflects them back toward Earth, mirroring the processes of perception and communication. It carries the same radioactive spark that probably liquefies water inside Pluto, where under slightly different circumstances life might blossom.

Nothing the robot sends back can be detected by the eye among the pre-dawn firs, even with a telescope. But something else is bouncing among the tiny Earth-bound telescope's mirrors that the robot can't detect. Signals of vastness, loneliness, and power reflected in as if from a sentinel on the edge of everything.

Pluto, though, to us is lifeless. A remnant of ancient warfares among planetary gravities. Whatever the robot finds will be frozen, dead matter. A corpse, in a way. Its very epinoian name reflects the ruler of the place we all descend to: dark oblivion, or the energy that coheres some eonic afterworld from whose bourn no traveler returns. The robot is configured to find dead matter only, and being no more than a reflection of a human being, can't imagine anything else. The human scientists are trained to dissect, analyze, and describe the findings. They hope to find inside the body an interior spark that—hope upon hope—will help us learn about ourselves. Inside Pluto is a wealth of information.

"Let us not underrate the value of a fact," Thoreau observed; "it will one day flower in a truth." He meant that the physical world is a reflection of nonsensory cosmic textures; and he thought of those textures as the moral world, which itself is a kind of foothills or border to ranges farther up the mountain where some mystics and enlightened scientists are taking a rest. They see that matter, a momentary configuration of energy, is essentially empty. If they followed the conventions of science they would become mired in dissecting details and find, to paraphrase an ancient mystic, nothing more than a corpse. The facts, isolated from the cosmic texture, are empty desolation. But when the faculty that detects beauty recognizes that emptiness, then silence blossoms into a clarifying voice, silently shining to silent moons.

New Horizons, unbeknownst to itself, has gone to the borderlands on a quest for gnosis.

Sources for "Pluto on the Borderlands"

Boyle, Alan. *The Case for Pluto.* Hoboken, NJ: John Wiley & Sons, Inc., 2010.

Bruno, Giordano. *On the Infinite Universe and Worlds.* Translated by Dorothea Waley Singer. New York: Henry Schuman, Inc., 1950. www.positiveatheism.org/hist/bruno00.htm#TOC

Davies, John. *Beyond Pluto.* Cambridge: Cambridge University Press, 2011.

Eddington, A.S. *Science and the Unseen World.* New York: MacMillan, 1929.

Gilster, Paul. "The Case for Pluto's Ocean." *Centauri Dreams* (Nov 2011): www .centauri-dreams.org/?p=20717.

—. "Pluto: Moons, Debris and New Horizons." *Centauri Dreams* (July 2012): www.centauri-dreams.org/?p=23638.

—. "Pushing Beyond Pluto." *Centauri Dreams* (May 2012): www.centauri-dreams .org/?p=22863.

Grossinger, Richard. *The Night Sky: The Science and Anthropology of the Stars and Planets.* Los Angeles: Jeremy P. Tarcher, Inc., 1988.

Jastrow, Robert. *God and the Astronomers.* Readers Library, 2000.

Jewitt, David. "Kuiper Belt." www2.ess.ucla.edu/~jewitt/kb.html.

Jones, Barrie W. *Pluto: Sentinel of the Outer Solar System.* Cambridge: Cambridge University Press, 2010.

Lakwadalla, Emily. "Pluto on the Eve of Exploration by New Horizons: Is there an ocean, or not?" *The Planetary Society* (Aug 2013): www.planetary.org/blogs /emily-lakdawalla/2013/08012202-plutosci-thursday.html.

Littmann, Mark. *Planets Beyond: Discovering the Outer Solar System.* Hoboken, NJ: John Wiley and Sons, Inc., 1988.

Lunine, Jonathan. "On Ammonia and Astrobiology." *Astrobiology Magazine* (Mar 2005): www.astrobio.net/index.php?option=com_retrospection&task=detail&id=1505.

Minor Planet Center. www.minorplanetcenter.net.

Mitton, Jacqueline. *A Concise Dictionary of Astronomy.* Oxford: Oxford University Press, 1991.

—. *Cambridge Illustrated Dictionary of Astronomy.* Cambridge: Cambridge University Press, 2007.

NASA. New Horizons website. http://pluto.jhuapl.edu.

"New Horizons Spots Pluto's Moon Charon." *Astrobiology Magazine* (Jul 2013): www .astrobio.net/pressrelease/5555/new-horizons-spots-plutos-moon-charon.

NASA. "Pluto Overview." solarsystem.nasa.gov/planets/profile.cfm?Object=Pluto.

"New Hubble Maps of Pluto Show Surface Changes." HubbleSite. Feb. 4, 2010. www.nasa.gov/mission_pages/hubble/science/pluto-20100204.html.

Olkin, C. B., et. al. "Pluto's Atmosphere Does Not Collapse." arxiv.org /abs/1309.0841.

Plato. *Cratylus* in *The Collected Dialogues of Plato Including the Letters.* Translated by Benjamin Jowett. Edited by Edith Hamilton and Huntington Cairns. Princeton: Princeton University Press, 1963.

Redd, Nola Taylor. "Pluto's Hidden Ocean." *Astrobiology Magazine* (Nov 2011): www.astrobio.net/exclusive/4343/plutos-hidden-ocean.

Robuchon, Guillaume, and Francis Nimmo. "Thermal evolution of Pluto and implications for surface tectonics and a subsurface ocean." *Icarus* (August 2011).

de Santillana, Giorgio, and Hertha von Dechend. *Hamlet's Mill: An Essay Investigating the Origins of Human Knowledge and its Transmission through Myth.* Boston: David R. Godine, 1977.

Schmude, Richard Jr. *Uranus, Neptune and Pluto and How to Observe Them.* New York: Springer, 2010.

Stern, Alan and Jacqueline Mitton. *Pluto and Charon.* Hoboken, NJ: John Wiley & Sons, Inc., 2005.

Thoreau, Henry David. "Natural History of Massachusetts." *Excursions.* Gloucester, MA: Peter Smith, 1975.

2

Pluto and the Kuiper Belt

Selections adapted from
The Night Sky: Soul and Cosmos

RICHARD GROSSINGER

In 1722, Johann Titius developed a Laplacian mathematical progression for the distances of the planets from the Sun. That same year, Johann Bode, in his *Introduction to the Study of the Starry Sky,* restated the progression with the specific recommendation that there was *another* planet at 2.8 times the Earth's distance from the Sun (or seven times Mercury's distance and four times Venus's distance). When Uranus, upon discovery, fit Bode's progression so closely, the search for Bode's "missing" planet was intensified, but it was not until New Year's Day, 1801, that Giuseppi Piazzi, an Italian astronomer, found an 8th-magnitude star in Taurus that changed position over the next two nights. On January 14, it ceased its retrograde motion and began moving forward in relation to the Earth's motion and the reciprocal positions of the two bodies. Assuming it to be a full-fledged planet, Piazzi named it Ceres. Then he lost it as Taurus passed into sunlight, after which he became ill and was unable to continue tracking. It was not located again until December, when Karl Friedrich Gauss, using an orbit computed from only three figures, spotted it.

Thereafter, astronomers flocked to see the new world. Ceres fit Bode's progression almost perfectly, but it was so tiny that it offered no planetary disk. While astronomers were puzzling over this, Wilhelm Olbers, in April of 1802, found a second object in a proximate orbit; this was named Pallas. Juno was detected nearby in 1804 and Vesta in 1807. By 1852, there were twenty known asteroids (or planetoids, as they were also called), and by 1870, there were 110. Once the chase was joined, they were disclosed all the way down to the 12th magnitude, including some that swing beyond Saturn and others that pass within the orbit of Venus. Hence the harmonic gap, the planetary (and musical) interval between Mars and Jupiter, like a shattered piano was filled provisionally by a multitude of tiny keys (planetoids) that also strayed from the soundboard. Of these, Hermes, Eros, and Icarus have each made near approaches to the Earth in recent times. We are not talking about meteor-size rocks. Icarus is almost a mile across and weighs 2.9 x 1032 kilograms. In New York City, that would stretch from 42nd Street to the Upper West Side in sheer spherical girth, which is the size of the crater it would leave while preserving little evidence of anything prior. It's a baby but a big baby.

In modern times we have come to appreciate the existential danger to our world from stray solar bodies of planetoid size, for an "impact event" with even a medium-size asteroid would release as much energy as several million nuclear bombs detonating simultaneously and might set Earth's oxygen on fire. Our most immediate threat (in 2013) is a thousand-foot-diameter asteroid named Apophis which presently stands a 1-in-250,000 chance of having its orbit knocked into a terrestrial collision course with an estimated arrival of April 13, 2026. In risk-assessment analysis, this may not be a high-probability hazard, but it is off the charts in magnitude of potential loss. The difficulty of implementing effective countermeasures for Earth's vulnerability to renegade worlds can *never* exceed the expectation of damage, hence the recent scurry to identify and track even the tiniest asteroids' orbits—there is always the danger of a Cosmic Hustler:

> Paul Newman (as Fast Eddie): "Fat man, you shoot a great game of pool."
> Jackie Gleason (as Minnesota Fats): "So do you, Fast Eddie."[1]

In 1846, John Adams, the British astronomer, and Urbain Leverrier of France simultaneously predicted the discovery of an eighth planet, using Newton's laws and computing its positions by the hypothesized perturbations that an unknown planet had caused in the positions of Uranus. Neptune was first observed, at least with awareness of what it was, by German astronomer Johann Galle, who used Leverrier's predicted position of September 23, 1846. After finding an 8th-magnitude star that was not in a current star chart he was using, Galle continued to observe it and recognized that it was moving in a retrograde direction in close proximity to Leverrier's estimates of size and speed as well as location. He then wrote to Leverrier that "the planet whose position you have pointed out *actually exists.*"[2]

On September 10 of that year, John Herschel peered through his telescope at this new world and declared: "We see *it* as Columbus saw America from the shores of Spain. Its movements have been felt trembling along the far-reaching line of our analysis with a certainty hardly inferior to ocular demonstration."[3]

It was like prying open the door to a mausoleum that had been sealed for a billion years.

A month later, William Lassell, using a new mirror telescope, discovered Triton, the very large moon of Neptune, which is brighter from Earth than any of Uranus's moons. It was named by Camille Flammarion in 1880 in *Astronomie Populaire,* but until 1933, it was more familiar to the world as "Neptune's satellite."

It is interesting to note that, even though a planet is discovered at a certain historical period, it is often recognized later that it was already familiar as a star. The ancients saw Uranus, but it was too faint and slow-moving to excite interest in its possible planetary nature. John Flamsteed observed the planet at least six times, the first in 1690, but he catalogued it only as 34 Tauri. Then between 1750 and 1769, French astronomer Pierre Lemonnier recorded Uranian sightings at least twelve times, four of them on consecutive nights, yet still considered it a star or several stars. American astronomer Sears Walker searched Lalande's eighteenth-century catalogue for unknown stars on the nights and in the positions where Neptune would have been visible, and he used these likely sightings in completing its improved elliptical orbit in 1847.

When Neptune was found to be well off its expected position from Bode's progression (30.05 times the distance of the Earth from the Sun instead of 38.4), the "law" lost any remaining credence. It actually belonged more to Kepler's universe of geometric solids, and its main role had been historical in guiding the search for new planets.

The forecast of Neptune by two independent experimental astronomers in the same year was not without controversy. Leverrier was the more famous and also the more egotistical of the two, and Adams's prior discovery went unreported for many months because the British Astronomer Royal, George Airy, did not believe in the existence of trans-Uranian planets and did not take his results seriously. When Adams's prediction was announced later, Leverrier and his French supporters vigorously fought against sharing the historical credit. In fact, Leverrier wanted the new planet given his own name and self-servingly suggested the name Herschel for Uranus.

This competition masks another, far more interesting complication that has been forgotten but not resolved. The credit for discovering a planet generally goes to the person whose formulas predicted its position, often instead of the person who actually finds it. Galle found Leverrier's planet for him. But Galle's "star" also corresponded to Adams's prediction for the same night, even though Adams and Leverrier had different orbits for their planets, neither of which corresponded even remotely to Neptune's actual orbit, nor to each other, except at the periods when they all intersected. That the three ellipses should have come so close to perfect intersection (the hypothetical ones of Adams and Leverrier and the real one of Neptune the planet) on the night that Galle chose to look is a stroke of synchronicity and luck almost too fantastic to believe.

Both sets of calculations were based on Bode's inaccurate estimates of Neptune's size and distance from the Sun. They were also based on the extremely subtle calculations of not only Uranus's perturbations of the unknown body but that body's perturbations of Uranus, and on an accurate Newtonian orbit for Uranus, with all of the other gravitational influences on it in the Solar System. This is a difficult "solution" to arrive at for many of the outer planets even today. Perhaps Neptune was found because so many astronomers were in the hunt. (By the same logic, the hypothetical planet that lies between Mercury and the Sun, Vulcan, was "observed" many times throughout the nineteenth century because so many astronomers were looking for a body to resolve anomalies in the perihelion of Mercury.)

Given these controversies, officials of the time chose to credit Leverrier with the discovery of Neptune because it was found through using his formula, *even though* his hypothetical planet and Neptune crossed at only a few points. Having bestowed this accolade, the officials were compelled later to give Adams equal credit, since his planet crossed both Leverrier's and Neptune at the discovery point.

A related dilemma became more significant in subsequent decades, after it became clear that neither the existence of Neptune nor the later discovery of Pluto fully resolved the anomalies in Uranus's orbit. Even today with advanced computers and adjustments from numerous Pluto-size Kuiper Belt Objects, we still can't find what is causing these perturbations—it is not trivial and it may not even be singular in source. Crypto-astronomers and conspiracy theorists have furnished astrophysicists with a gaggle of mythical planets, imaginary planets, hidden planets, and artificial planets—"Nibirus" and "Nemeses," anti-Earths and mirror Earths—for mainstream scientists (of course) to summarily dismiss. But these worlds' ghosts and collective menace haunt the outskirts of solar cosmology. We just don't know all that's out there, buried in long orbits or camouflaged fields. For our substantial hubris, we are quite insular and sequestered in our own backyard.

The Neptune situation is summarized well by the historian William Graves Hoyt, quoting from the American astronomer Henry Norris Russell: "Leverrier and Adams . . . had assumed too great a distance for their unknown planet from Bode's law, but their calculations also 'made the orbit considerably eccentric (although it is really very nearly circular),'" and this "spurious eccentricity brought the predicted orbit toward the sun in the region where the planet actually lay at the time, and went far to undo the error of the two original assumptions."[4] Apparently Neptune was ready to be found.

Untold other planets may exist between Neptune and the comets, invisible to us for the same reason that planets circling other suns are invisible: their faint reflected light does not reach us. Their sizes would be limited to a range at which their gravitational effects would not be readily apparent in the rest of the Solar System. Phantom planets received far more attention early in the twentieth century than they do now, as astronomers regularly extrapolated them from unexplained perturbations in the orbits of Neptune, Uranus, and even Saturn, and from anomalies in orbits of comets.

William Pickering, the American astronomer at Harvard, was the most prolific planetary model-maker between the years 1928 and 1931. His first quasi-planet, "O," was placed at almost fifty-two times the Earth's distance from the Sun. It orbited the Sun every 373.5 years and had a mass twice that of the Earth's. Pickering derived "O" in 1928 solely from residuals in the orbit of Uranus. A few years later, he announced three more planets: "P" was located at 123 AUs, with a year of 1,400 Earth-years; it was based on variations in cometary orbits. "Q" was located at 875 AUs; it had a mass twenty-thousand times that of the Earth, a year of twenty-six thousand Earth-years, an eccentric orbit, and a sharp orbital inclination. "Q" was big enough to counterbalance the whole system—to operate, gravitationally, as a second, invisible Sun. "R" was located at 6,250 times the Earth's distance from the Sun. Its mass was half that of "Q," and its orbital period was half a million years.[5]

"S" was located at forty-eight times the Earth's distance from the Sun, with five times the Earth's mass and a year of 336 Earth-years. "T" was invented to cover the few remaining anomalies in Uranus's orbit and was placed just outside the orbit of Neptune. "U" was later placed outside the orbit of Jupiter in an eccentric path that sometimes carried it closer to the Sun than the giant planet; it was only 4.5 percent of the Earth's mass.[6]

Needless to say, none of Pickering's planets has ever been found.

The quest for a ninth planet in our Solar System began in the manner of the search that led to the discovery of Neptune. It was an exhausting, fruitless obsession over three decades, with an ambiguous resolution in 1930. Percival Lowell began the quest at the turn of the century from his observatory at Flagstaff, Arizona. Working with unexplained residuals in the orbits of Uranus and Neptune, he calculated the probable orbit of Planet X, and then he and his staff proceeded to search for it throughout the likely regions of the sky. But Lowell recognized "that when an unknown is so far removed relatively from the planet it perturbs, precise prediction of its place does not seem to be possible. A general direction alone is predictable."[7] He died in 1916, before X's discovery.

The man who would eventually "land" Planet X was Clyde Tombaugh, a self-taught astronomer who worked on his family farm in Kansas during the day and studied the heavens at night with a nine-inch reflecting telescope that he had constructed himself. He applied repeatedly to the Lowell Observatory for a job; astronomer Vesto Slipher, impressed with his drawings of Jupiter ("fairly good for such a chap working all alone")[8] eventually hired him. Slipher gambled that this ambitious, self-stoked lad with only a high school education would be more productive in the search for X than another university type. Tombaugh was twenty-three years old when he became the last person to discover a new solar "planet."

Tombaugh worked by exposing plates on good nights and then "blinking" successive exposures of the same area on different nights in a special machine to see if any "star" moved. Searching down to the 17th magnitude, he encountered innumerable difficulties in both exposing the plates and producing blinkable pairs. The most serious obstacle was the thousands of asteroids moving through the zodiac. Tracking at this high level of magnitude, he was plagued by their constant blinking. "The key strategy," Tombaugh concluded, "was to take the plates within 15 degrees of the opposition point, so that the daily parallactic motion of the Earth would cause all bodies external to the Earth's orbit to retrograde through the star field—the more distant the object the smaller the angular displacement during the interval between the dates the plates were taken. This solved the asteroid problem beautifully."[9]

After months of scouring Lowell's proposed orbit for Planet X without a planet-like glimmer, Tombaugh began an exhaustive search of the entire zodiac in September 1929. He snaked from Aquarius into Pisces, from Pisces into Aries, and then into Taurus and Gemini. In eastern Taurus and western Gemini, the average number of stars per plate increased from fifty thousand to five hundred thousand as he crossed toward Milky Way center. By early November, Tombaugh was in Scorpius and Sagittarius and was encountering a million stars per plate. Disappointed, he returned to western Gemini in January 1930. Blinking two good exposures of the Delta Geminorum region from January 23 and 29, he saw a shifting star, Planet X.

Since it was found at Lowell's observatory, under the auspices of his search, by an astronomer using his tables, Lowell was given credit for having discovered the planet along with Tombaugh. Meanwhile, William Pickering claimed that the new body was his Planet "O"; then, when he learned of its meager size, he discarded it immediately as a significant perturbing influence on its neighbors. The perturbation on Neptune's orbit requires a planet or medley of planets with diameters adding up to something in the neighborhood of fifty thousand cubic miles' worth of gravitational pull. Pluto's girth weighs in, at most, at 4,200 miles, while some astronomers believed at the time that it was no bigger than Ceres (around six hundred).

In any case, Pluto turns out not to be the *last* of the planets (as presumed for decades) but the *first* of the Kuiper Belt Objects. As a KBO or dwarf planet, it doesn't so much close the legendary curtain on Uranus and Neptune as open a new one on Varuna and Quaoar.

Astronomers conducting photometric research on Pluto late in the twentieth century at the University of Hawaii concluded that Pluto's mass was a few thousandths that of the Earth. This is "much less than would be required to perturb the motions of Uranus and Neptune measurably. . . . If this train of logic is basically correct, it appears that Tombaugh's discovery of Pluto in 1930 was the result of the comprehensiveness of the search rather than the prediction from planetary dynamics."[10]

But then, where is Lowell's predicted trans-Neptunian planet? Forget Nibiru for now—that is a different myth. *Something is still missing.*

For all intents and purposes, Pluto had been used to save the appearance of Lowell's phantom, which it didn't. It did, however, give mid-twentieth-century astronomers a way to pretend that the mystery had been solved (more or less), much as nineteenth-century astronomers pretended that Ceres had saved Bode's law.

Even the choice of its name, Pluto instead of the more popular Minerva, conferred a tinge of murk and spooky fortuity, not only that the overseer of the Roman Underworld should rightly rule a remote, sunless realm but that the first two letters of the name of a major available god should be Percival Lowell's initials. The PL symbol for the planet thus could indirectly honor Lowell. Yet clearly it is no more his planet than Pickering's, though Pickering was quite satisfied with the name, which he read as Pickering-Lowell.

A Plutonian satellite, named Charon, was discovered by James Christy and Anthony Hewitt in 1978. This allowed measurement of Pluto's mass by Kepler's third law and confirmed the small size of the planet and the relative largeness of its moon (1,430–1,443 miles in diameter for Pluto, 750 for Charon being the most recent estimates at the time of this writing). Charon and Pluto revolve around each other every 6.387 days. Pluto/Charon turns out to have four other tiny moonlets: Nix, Hydra, Kerberos, and Styx. The latter two were found in 2011 and 2012, respectively.

Plutonian moons' approximate diameters are loosely derived from ranges suggested by their geometric albedo. Nix falls somewhere between 28.5 and 85 miles in diameter (depending on its actual albedo) with an orbital period of 25 days. Hydra is between 38 and 104 miles in diameter with an orbital period of 38.2 days. Nix is in close to 1:4 orbital resonance with Charon, and Hydra close to 1:6 with the larger moon. Kerberos, between eight and twenty-one miles in diameter, is itself in 1:5 orbital resonance with Charon, while Styx, six to sixteen miles across, shares a 1:3:4:5:6 sequence of near resonances with Nix, Kerberos, and Hydra. The barycenter of their orbits does not lie within any single body, so "Pluto" is more properly a swarm. It is not so much that one relatively large moon and several smaller ones orbit the planet as that the entire swarm orbits its own center of mass.

Prior to the discovery of Charon, astronomers favoring Pluto as the perturbing influence on Uranus and Neptune offered a number of mostly bizarre explanations for its ostensible small size. Some hypothesized that it is such a dark frozen planet that we see only a three-thousand-mile-wide glint of its fifty-thousand-mile-wide sphere, a sparkle on a shiny ebony ball. That led to another, only slightly more likely but decidedly more titillating suggestion: that Pluto is a white dwarf or a black hole, a burned-out cinder from the other half of a twin Sun system of which our present Sol is the remaining representative. Pickering jumped on that bandwagon briefly, hoping to salvage his Planet "O" by offering that it might be a stray white dwarf picked up from another star system and thus the only body in the Solar System with an origin outside it. These attributes would give it the necessary mass for its supposed perturbations.

With more accurate measurements of the body's mass, assorted Plutonian fantasias have fallen by the wayside, though the mystery of the secret perturber remains. If not Nibiru, perhaps it is the cinder-sun Nemesis (which is said to approach its rival Sol every twenty-six million years with a glorious retinue of comets). (I am playing dumb astronomically but not mythologically.)

In endless night, Pluto travels 2.7 billion miles from its governing star, at which remove it takes 248 Earth-years to make its annual circuit. We know how fast light travels: 186,000 miles per second is beyond conception, faster than a cartoon roadrunner. Blink and it's traveled from New York to Paris and back twenty-five times. Yet, the Sun's light still requires 3.22 real-time minutes to reach Mercury. A twinkle of that same ray takes five hours and twenty-eight minutes to reach Pluto. If we imagine a beam of light whipping off the Sun's surface at 186,000 miles per second, zooming down the corridors of interminable darkness and space so fast that it opens shafts that disappear into the distance even as they come from a remote darkness in another direction, then we can understand how far is that five-hour-and-twenty-eight-minute light-journey.

The temperature at Pluto, minus 369 degrees Fahrenheit or lower, would freeze any atmosphere. If a flock of magpies were released there, they would shatter against the sky faster than they would congeal on a bare branch in a Saskatchewan winter.

As a child of the primal nebula, Pluto's composition should be water, ammonia, and methane ice, with occasional crisp haze, for these are the last substances to condense from the disk's outer rings in the cold of deep space. A plausible planetary model calls for a water-ice core, a methane and ammonia mantle, a frozen methane crust, perhaps some dead methane seas, and, at most, a scant atmospheric presence. There is a remote chance of an underground sea, perhaps warmed to liquid by geological activity. What a strange lagoon that would be! Imagine traveling a couple of billion miles across frigid, black space, landing on gaseous metallic rock, and then taking a hydraulic lift a few thousand additional miles through a tunnel to bathe in a sacred Plutonian hot spring.

The world briefly known as Minerva is almost certainly a barren, spaless KBO.

Because Pluto is the only full-fledged orb traveling outside the ecliptic that makes up the zodiac—i.e., transiting at an angle to the plane of the Solar System like a sewing needle and thread darning into and out of a piece of fabric, the fabric containing the other planets and moons—it was presumed for its first sixty years most likely to have been a moon of Neptune that got knocked loose by a large scattering object. Recent computer models, however, have indicated that, no matter the scale of an intruder, gravity and debris conditions could not have displaced anything the magnitude of Pluto from Neptune's attraction and orbit into its present position. Most scientists now think Pluto formed in the early Solar System through a sort of checkered accretion and crashing process.

The dwarf planet's course around the Sun is also quite eccentric compared to Neptune's near-perfect circle. In fact, on December 11, 1978, Pluto moved within Neptunian orbit, reaching perihelion on September 12, 1989. Neptune then became the outermost planet of the original nine and remained so until February 11, 1999, when Pluto swung out beyond it again, en route to aphelion in 2113. In about 230 years Pluto and Neptune will switch places again. At the same time, Pluto stands in 2:3 orbital resonance with the outermost ice giant.

A Plutonian day is hard to fix without surface features, but variations in brightness suggest 153 hours. Pluto's density is guessed to be 88 percent that of the Earth, in this respect more like the inner than the outer planets.

It will have taken 9.5 years for NASA's *New Horizons* spacecraft, launched January 19, 2006, to escape Earth's gravity well and be slingshot by gas and ice giants to Pluto. When it arrives at the swarm on July 14, 2015, I would like it to photograph boulevards of apartment buildings and shops resembling midtown Manhattan. I look forward to the face-saving explanations by scientists. I would savor the reductionist interpretation of the cosmos failing at the edge of our Solar System, as Pluto directs our attention outward to Centauri, not back to Terra, for the lost continent and its archetypal city.

♇

Pluto/Charon at the traditional Solar System's outer gate, clandestine in its orbit, bears death and rebirth, absolute power and "soul intent." Unity Existence put its nano-hand on Pluto/Charon on February 18, 1930, bringing its blip into worldview at precisely the moment that its manifestation was poised to flood over us. But nothing actual was transmitted—no liqueur, radiation, or ether spilled on the Earth. Springs were loaded. They uncoiled forces accessible only in horoscopes, as they potentiated a psycho-astronomical field. Whole universes of being sprang into expression under Pluto, as inner worlds turned over, casting their debris outward. Unsignified meanings never before on Earth broke through.

Of course, Pluto itself is a nubbin of a world, demoted from even full planethood. Clearly astrologers are not talking about the gravitational tug of the swarm: a pip with an accompanying moon and a few moonlets. Instead, they are reading an occult clock that uses a Plutonian marker to designate a position and timing mechanism in an otherwise imperceptible field.

The series of near resonances among Pluto's moons could be used, at least symbolically, to explain the amplification of the planet's astrological transmission to the rest of the Solar System. Of course, to have such a pending explanation you would have to know how astrology "works," so it's best just to say that satellite resonance coincides with the sign's augmentation.

Pluto probably hatched forms astrologically on the zodiacal bodies of Neptune, Saturn, Jupiter, Ceres, Chiron, Mars, Venus, and all the other proximal worlds and objects insofar as each has a potential alias within terrestrial astrology as well as an astrological field of its own. At the moment that it entered Earth's symbolic field, Pluto pervaded every world in the Solar System on subtle planes, though it was already operating universally at an unconscious, nonhuman level. Call it a metaphor for the diversity and depth of Plutonian influence and its ensuing terrestrial projections onto other planets and planetoids after its discovery but, in any case, the relationship between Tombaugh's and Pickering's hidden perturber and the symbols it instantaneously generated made the signs interchangeable and correlative; after all, *every single one worked.* They probably would have worked even if the Lilliputian orb had been Minerva.

Pluto is a realization of dark forces that have been operating under the radar and then suddenly burst into cognizance. This is what happened after the planet's identification in Gemini and its subsequent promulgation and christening. Once Pluto was accepted into the human gestalt, Plutonian forces began inculcating the Earth. Skeptics say that it's the other way around: astrologers merely gave the new planet some of the more current and endemic attributes of our changing world. Yet under synchronicity, *these are the same thing.*

Pluto is a camouflaged planet, an uncertainty planet, a placeholder for black holes and artificial worlds. Its coyotl-like arrival in one of Tombaugh's blinks was enigmatic in a precisely Plutonian way. From its long, eccentric orbit it continues to rule metamorphosis, regeneration, resurrection, reincarnation, and magnetism.

While Uranus marks the onset of atomic power in the natural element uranium, and man-made neptunium follows on the elemental chart, the most dangerous of the elements released by man into the atmosphere, oceans, and life cycles is plutonium. It is the custodian of the discovery of atomic fission on Earth. Pluto dispenses earthquakes, tsunamis, epidemics, world wars, nuclear accidents, and pole shifts. Its magicians are fakirs and yogis. Its honchos are sociopaths, warlords, terrorists, Mengeles.

From its Underworld association, Pluto bears the horror and gloom of images that are so deeply and collectively unconscious that we cannot get at them, or else we receive them only as nightmares, shapeless and shadelike, akin to death itself.[11] The lord of the Underworld travels the heavens with his boatman Charon. "With his queen Persephone he held sway over the other powers of the infernal regions, and over the ghosts of the dead. The symbol of his invisible empire was the helmet that made men invisible."[12] Psychically, Pluto *is* a black hole.

Once the grim ritual behind the Nazi swastika was disclosed, astrologers diagnosed Hitler's reign as an epitome of negative Plutonian influences coming to bear on Earth following Tombaugh's revelation, i.e., horrific aspects that had long been buried beneath human consciousness.[13] The Nazi ritual was so inhuman and weird that it might well have come from Pluto or the uncharted domain that astrological Pluto rules. It is where Aryan heavyweight Odin meets and conspires with the alien Greys and Arcturians of UFO lore.

Some occultists believe that, of all the massacres, genocides, and mass slaughters on Earth, the Nazis' transgressions alone *were channeled from extraterrestrials.* Hannah Arendt's recognition of "the banality of evil," her umbrella rationale for the Holocaust, carries a vague subtext of off-planet supervision and choreography (not to her, of course, but to me): the Holocaust is banal because it doesn't support its own ritual; it doesn't match or justify its output; it doesn't fit or belong here. Why Germans of the Weimar era performed such a ceremony we will never know, but the trance in which they acted speaks to a hidden perturber or some other Kuiper Belt archetype.

Pluto is mythologically the source of wealth and grain, each of which comes originally from the earth. His allies are amphibious animals, poison-bearers, ants, spiders, and hairy plants. Black diamonds and deep purple-blue stones—or anything undergoing metamorphosis (cocoons, stalactites, stones changing color)—are Plutonian.[14]

Pluto's cruelty and inexorability are part of the cosmic order, but he is a cutting-edge paradigm shifter too. An eccentric planet, slightly out of the zodiac, passing slowly through each of its signs, he is the perfect agent for millennial and cataclysmic changes. Astrologer Ellias Lonsdale adds:

> The force of Pluto is like a force from a vast vast vast desert way way way out there. Pluto is there to draw us into a greater universe. Like it's the last outpost that keeps reinstilling us with a cosmic perspective and keeps taking away our materialistic assumptions.[15]

According to psychologist James Hillman, Pluto's real message is neither Hitler nor spiders nor the planetwide deployment of atomic power, though these are Plutonian factors. It won't be a mind-matter ray or psychic electricity either—passing Plutonian manifestations. Pluto is the Great Abyss itself: a vacancy, a lesion, a latency so deep that it cannot manifest as form in this world-age or even this cosmos. The planet is missing from its own sky, as the Olympian deity is pre-erased from Greek art—no figuration or likeness of him even where he is known to abide. Pluto is an adumbration incapable of congealing, a fathomless shadow, a homunculus sealed behind its own cenotaph. It is mystery of Nature itself, of existence and beingness. Pluto is the god of existentialism, the paradox of his own deletion, millennia before Sartre or Derrida.

Hillman adds:

> All this "negative" evidence does coalesce to form a definite image of a void, an interiority or depth that is unknown but nameable, there and felt even if not seen. Hades is not an absence, but a hidden presence—even an invisible fullness.[16]

The despair of Pluto is inescapable, though we have set up our entire culture to evade it. In fact the prime motive of most terrestrial lives is to avoid and eradicate Pluto: have fun, be social, amass possessions, make small talk, fuss over houses and cars and parties—in sum, to pretend Pluto doesn't exist. And it didn't, for a long time.

Our terrain has to shift because it is ephemeral and artificial. Before Pluto arrived, we inhabited a consensus landscape, waging the narrative of a life through which we flattered and consoled ourselves. This was as true in ancient Persia and Egypt as it was during the Middle Ages and the Renaissance. It came into its own with the industrial revolution. At heart, we didn't believe any of it. We didn't even really want it. Like Pluto, nothing was actually there.

But don't blame Pluto for our banality or brutality. Pluto is simply absence. Absence gets filled, whether by sourceless ceremonies or various actual and symbolic punishments and prisons, by bottomless grief and everyday neurosis. Pick your level because Pluto is also *sunyata,* a Sanskrit term for lack of inherent essence but also spaciousness, openness, wisdom, enlightenment, or, more precisely, the clarity at their intersection.

When we come into contact or conjunction with Pluto in its 246.04–year orbit dwarfing a human lifetime, we experience the full, stark intensity of our situation and dilemma, the prodigious emptiness and futility at the heart of existence. In Pluto, there is no hope, no possibility, no chance that we will ever feel better. Pluto is terminal depression, the perdition of everything that reassures us. It is that barren a world, that far from the Sun, that much in exile from the current hustings of the Solar System—and I don't mean just astrophysically.

All tiny, dark worlds leave mute and invisible footprints on our doorsteps and inside our hearths. They can be as negative as Nazi gas chambers or as positive as the Resurrection of Christ. Under Pluto, it's all the same because it is doing the Great Work, only in 246–year plumbs—world enough and time to swing a pendulum between Christianity and fascism more than once, to ride out any history or meaning.

The way to come out of Plutonian sorrow is to depart it by an entirely different gate from the one by which you entered—as a different person. And that is not only straightforward and available: it can be done without any drama or fuss *as soon as one is willing.* Passing through Pluto's reality means a consciousness shift, a transit to unfamiliar ground. What is habitual and unconsciously reassuring can no longer exist on Pluto, for Pluto alters outright, and the only hope is to *become* its unknown thing.

Resurrection is the point of Pluto anyway. It completes its cycle and takes off its yoke only when recognition of its sheer *difference* becomes more powerful than compulsive preoccupation with its emptiness.

The moral: the ground we stand on is neither recreational nor social nor, for that matter, does it have anything to do with who we tell ourselves we are. Those comprise the inner planets: fleeting solaces, symptomatic stages for dramas as we bide our time and await determination.

Pluto gets to us and deepens us by wiping that slate clean, by locating us where we are actually standing, where we were originally rooted. It continually finds us pretending and makes us real. As it darns its way stealthily in and out of Neptune's orbit, it periodically touches bottom. It says: "What seems to be happening couldn't be happening." And it isn't.

Plutinos and Other Trans-Neptunian Planets

Pluto was a fully qualified if peewee planet until astronomers realized that it belonged to a class of smaller trans-Neptunian objects that orbit the Sun at Pluto's distance and beyond. The celestial mechanic didn't carefully line up four gas and ice giants after four rocky planets (and rock-strewn asteroid belt) only to cap them with a *single* rocky body of equal status yet smaller than either Mercury or Mars. Pluto could stand as the outermost planet only as long as we were blind to a string of comparable worlds orbiting beyond it.

Many folks have since lamented the demise of Pluto's planetary status, for we had gotten used to its pup-like yelp at the end of the solar roll call. Yet this was more street mythology than astronomy, with a bit of Walt Disney thrown in. There was even a brief dispute in the late twentieth century concerning whether the planet was named after Mickey Mouse's dog rather than the monarch of the Underworld, but Disney animators confirmed that the planet's naming on March 23, 1930, came several months before their rechristening of Rover (considered too common a brand at the time by company marketers). Pluto the pooch was not consciously Lowellian either; it just popped into their minds, but it gave a slightly daffy closure to the Solar System, an antidote to the somber Roman god. Who would kick a cute mutt out of a nursery rhyme anyway? Yes, the names of the planets are kind of a kindergarten pentameter, so the Solar System without Pluto seemed shorn of its last beat, its tail.

Obviously it is more elegant to group trans-Neptunian dwarfs together, for there are many similar-sized homunculi functioning as a second, back-woods lineup around Sol, most of them with their own moons, and each taking 247 years or longer to complete a solar orbit. Some of the worlds are even bigger than Pluto, another strike against granting it the closing slot in a sequence beginning with Mercury, Venus, Earth, and Mars. Yet Pluto's demotion was not only a matter of size, as Mercury and Mars would be relegated to dwarf status if they orbited the Sun beyond Neptune. It is size plus position plus historical context.

If the dwarfs were all included in a planetary census, the Solar System would lose its concision as well as its traditional pantheon of deities. The dwarf worlds simply don't belong at the same party as the rest. They don't come from Sumeria or Greece or Rome, or even Galileo's Italy; they are citizens of modernity: small digitalized rocks.

In an email to me, poet Robert Kelly remarked that it was not the demotion of Pluto *per se* that concerned him as much as the hasty coronation of *all* the dwarfs, giving astrologers license to clutter their already-muddled charts (per asteroids, centaurs, and Plutomania) with fresh binges of mismatched fancies and fads, forfeiting the singularity and holism of the horoscope by aggrandizing newcomers while depreciating pivots of more central and immediate orbs:

> My anxiety about the poor planetoid was funded more by a kind
> of wry remembrance of how the astrologers of the 1940s–1970s
> made such an important issue of Pluto in their reading of charts
> and their prognostications of new Eras of the world, etc. You
> remember. Now this new hierarchy, as you nicely call it, will
> invite a county fair of stargazing speculations, and we'll soon see
> horoscopes thick with those little fellers and their moons. And
> maybe in fact they'll get it all clearer than it has been. But I'm
> not sure how much the Sun and our Moon (our masters) care
> about them.

Yet conversely, if Pluto were made an exception and kept the ninth planet to the exclusion of the other KBOs, it would be favoritism *sans* astronomical context.

In addition, there is more than one "Pluto" (or plutino) traveling in Pluto's orbital slot. A plutino is a KBO locked in a 2:3 mean-motion resonance with

Neptune. For every two solar orbits that a plutino makes, Neptune makes three. So Pluto is a member of a gang that has been recruited by Neptune over the years using 2:3, one of the strongest and most stable resonances in a universe run by celestial mechanics. The Plutino groove is an attractive ride!

Sixteen plutinos are bright enough to classify individually, though only three of them have been deeded Earth names to go with their identification numbers. Beyond the sixteen, countless others that are more moonlet-like or meteoroid than planetary don't get serialized. In the long watches of astronomical time, Pluto has had to be careful not to bump into any of its co-orbiters, or maybe it hasn't and that is why it travels outside the plane of the Solar System with so many small co-moons.

Imagine another Earth and Luna orbiting the Sun at approximately 92.9 million miles but always on the other side of the ellipse. Numerous science-fiction writers have plotted just such an arrangement. Anti-Earth could be entirely different from Earth, a world of randomly drifting geological effects across a lifeless tundra, hence a candidate for human colonization. But it could also bear its own life forms, from a different biology and genetic molecule. It could be populated by humanoid tribes, speaking foreign languages, with different belief systems from us, and at their own stages of development. The inhabitants could be either more or less advanced than concurrent Gaians, arriving here in alien spacecraft or being visited first. They could have martial traditions and imperial ambitions and come as conquistadors, but so could we. Or they could be scientists and diplomats. Of course, that is the dichotomy we imposed on our other mythical twin, Mars, so it represents our ambivalence about our own essential nature as well as life in the universe in general: a fallen angel or a beast with a territorial imperative.

A few authors have tried to make Earth and Anti-Earth invisible to each other by placing the Sun always between us. Disclosure of the latter's existence would come as a shock to us at some point in the future. I'm not sure if that siting works geometrically, and it certainly doesn't gravitationally. At least two authors (that I know of) didn't even bother to put the Sun between us. They just had us be mutually unaware of each other's presence for no reason other than the convenience of their stories. In a 2011 movie based loosely on the latter trope *(Another Earth)*, Anti-Earth is unexplained either orbitally or astronomically; it just arrives. The worlds then turn out to be identical to each other in every way: they have all the same people and situations, until the moment that their inhabitants learn of the others' existence. From that point on, identical people on the two planets begin to deviate in small ways. So if you were responsible for something terrible happening on your "Earth," you could travel to the other and try to prevent or redeem it.

In any case, NASA's lunar missions established definitively, by changing the viewing angle, that there are no other "Gaias." Anti-Earth remains a mythological stand-in, though, for alternate realities.

Most dwarf planets are not plutinos; they lie beyond the Pluto cluster orbiting the Sun at a mean 247.3 Earth-years. There are too many of these trans-Neptunian objects (more than forty notable ones by 2014) for me to cover them all—and new ones are still being culled from the Kuiper Belt—so I will offer vignettes of only the major dwarfs, starting with the plutinos (though Kelly's warning is a worthwhile one: do not snag or litter a chart with too many remote planets and influences; these crest beyond the range of most individuals, at least in this incarnation).

Furthermore, like Pluto itself and unlike the interior Solar System, KBOs can have only transpersonal and karmic astrologies; they cannot leverage their remote pulleys at a fulcrum of mundane personality or relationship. It would be like trying to use a crowbar in New York to lift a rock in Beijing. They do, however, tunnel under entire civilizations, pebble by pebble. Being far from the Sun, they move very slowly relative to Earth, completing their orbits in spans well beyond ordinary personal situations or even cohesive historical periods. Two hundred and forty-eight years ago Mozart was performing operas and Britain was levying taxes on the thirteen colonies. We can tell the dwarf planets' meta-stories, but we can't get at them otherwise. Our mortality is too brief.

KBOs like asteroids tend to elucidate the astrologies of inner worlds. New to our zodiac, they are still depositing sigils and symbols, which are neither as positive nor negative as they might seem in a first pass. Like Pluto, other KBOs just take a long angle on human affairs. (Note too that the reign of solely European Graeco-Roman names ended at Pluto; the outer Solar System is fully multicultural.)

Orcus

Orcus was discovered by astronomers Chad Trujillo and Mike Brown at the California Institute of Technology along with David Rabinowitz of Yale

University on February 17, 2004, but appeared retroactively on photographs dug up from 1951. Many KBO dwarfs similarly hid, unnoticed, on earlier plates. Orcus is named after the Etruscan god of the dead. With an orbit like Pluto's of 245.18 Earth-years, this plutino (the purest anti-Pluto) is at aphelion when Pluto is at perihelion and *vice versa*. Pluto and Orcus are mirrors reflecting each other across the ecliptic. Orcus has a large moon emulating Pluto's Charon, which was named Vanth after the winged psychopomp who guides the souls of the dead to the Underworld, the Etruscan deity most closely associated with the Greek ferryman of Hades. Orcus appears bright gray, indicating a good deal of frozen cryovolcanic water mixed with methane and ammonia. Because of Vanth, scientists were able to estimate the parent body's mass as approximately that of the Saturnian moon Tethys. Orcus has a diameter of five hundred miles, Vanth about 165 miles.

In respect to their near-common orbits as they resonate astrologically, plutinos transmit a collective meaning set. Orcus specifies our perilous journey to the depths of our being, beneath the collective unconscious, beyond even the outlands represented by Pluto. It is where we will someday find an antidote to Pluto in an actual, nonsymbolic underworld, so it is where we encounter why and how beingness itself moved out of Eternity into Time. Orcus is the ultimate purifier.

Ixion

Ixion is a plutino too—its orbital period is 249.95 years. Redder and brighter than most KBOs, it passes below the ecliptic at perihelion as Pluto's perihelion passes above it, so the two worlds don't encounter each other catastrophically. In astrology it governs matters of tyranny, confrontation, and particularly "malicious and deceptive power-plays between authoritarian figureheads, drawn out in long karmic-based scenarios."[17] Think Saddam Hussein and the Bush family; the Irish Republican Army and the Protestant Orange Order; the Netanyahus and Yassir Arafat's Palestine Liberation Organization; Kim Il-sung, Kim Jong-il, and Kim Jong-un in succession facing down the ruling parties of China, South Korea, and the United States. Each side gains its strength and obstinacy from the placement of Ixion in the heavens.

Based in institutional authority rather than spirit, Ixion champions allopathic medicine's artificial prolongation of life over homeopathy's stimulation of the life force. Ixion wants to keep the old order in power at the expense of what must come; it is the plutino that most fervently tries to extend materialism's authority past materialism's capacity to hold a prerogative itself.

Huya

The biggest, brightest KBO next to Pluto at the time of its discovery in 2000, Huya turned out to be only about 272 miles in diameter, a roundish icy rock with a few bright dimples and an orbital period of 247.72 Earth-years. Identified by Colombian astronomer Ignacio Ferrin, it was named after the rain and winter god of the Wayuu peoples of northern Colombia and northwest Venezuela—a spirit that dwells beyond the Sun.

Astrologically Huya pushes people to extremes and, like Pluto, doesn't care whether the deed is honorable or shameful. He is quick-witted, shrewd, and as stubborn as Ixion, yet much more risk-taking and generous, especially when generosity serves his purpose. Huya's curious nature delves compulsively into how stuff works down to minutiae. In sum, he oversees the transmutation of fixed patterns from past lives into freer and more flexible expressions each next time around. According to online KBO astrologers, actress/host Oprah Winfrey incarnated a high dose of Huya energy: motivation of others, capacity for inner and outer change, the karma of the rainmaker.

Varuna

Varuna, a 643-miles-diameter trans-Neptunian object unperturbed by Neptune, was discovered by Robert S. McMillan at the University of Arizona in 2000; it orbits the Sun at 43 AUs in a near-circular course with a period of 283 Earth-years. Named after the Hindu deity of the oceanic and heavenly waters and the guardian of immortality, Varuna oversees our eventual return to cosmic law and the sacredness of life.

Varuna is also a sublimation of Jupiter, translating Jovian satiety and jollity into a kind of transcendental generosity with occasional relapses into megalomaniac narcissism. Alternately beneficent and malign, Varuna participated in the Parisian peace treaty ratifying the existence of the United States (conjunct the Sun, 1773), the assassination of President John F. Kennedy (transiting Pluto in exact opposition to natal Varuna, 1963), and the recognition of AIDS (conjunct the Sun, 1981).[18]

Pluto and the Kuiper Belt

Haumea

Haumea was detected by team Brown in 2005 and named by him after the Hawaiian goddess of childbirth and fertility who had pieces torn off her body; these developed into her children: fire lady Pele, sea goddess Namaka, and Hi'iaka, the deity of hula dancers. Haumea's orbital period is 285 Earth-years, just beyond Varuna. Assisted by Makemake, she is perhaps the most karmic of the KBOs, heralding the tiny chunk of our civilization that will survive the current epoch and beget the next phase of consciousness on Earth. As such, she governs the Hawaiian Sovereignty Movement, the break-up of the United States, the regaining of Texas, New Mexico, Arizona, and California by Mexico, and the kahuna principle behind the election of Barack Obama (though not his presidency).

About seven hundred miles in diameter and one of the weirdest objects in the Solar System, astronomical Haumea is a solid irregular rock with a thin glaze of water ice. Unlike other KBOs, it is ellipsoidally elongated like a rugby football, its major axis twice as long as its minor one. It also rotates very rapidly and has a higher-than-expected density and bright water-ice albedo.

Haumea sports two moons named after Hawaiian goddesses: Hi'iaka (217 miles in diameter with an orbit of forty-nine days) and Namaka (105 miles in diameter with an orbital period of eighteen days). This outlier dwarf is presumed to be the largest sheared-off survivor of a Kuiper Belt collision, which likely fissioned its moons too, as there is little surplus debris that far out in the Solar System. Though the impact occurred billions of years ago, Hi'iaka continues to perturb the motion of Namaka, the two moons oscillating on the edge of an 8:3 resonance, alternately getting caught and then escaping it over the aeons.

Quaoar

Quaoar is a rocky KBO, spotted in the constellation Ophiuchus on June 4, 2002, by Brown and Trujillo. It appeared on a Palomar Observatory plate in 1954 but was not identified as a "planet." It has one moon and, similarly to Varuna, orbits in a near-circular path just beyond Varuna and Haumea at roughly the same 43 AUs with little axial inclination and a period of 286 Earth-years.

Quaoar is named for the creation god of the Tongva, the native people of Los Angeles (from which the world was sighted). At the time it was the largest novel "planet" since Pluto (six hundred to eight hundred miles in

83

diameter, about one-third Pluto's size), but it has since been surpassed by Eris, Sedna, Haumea, and Makemake. Astrologically Quaoar regulates esoteric teachings from outside human consciousness as well as the coming shift in our relationship to the cosmos and our own egohood: away from individuality and toward multipersonhood and group souls. It is an attendant of indigo children, spiritually advanced reincarnates who manifest on Earth from elsewhere, often as autistic sages or idiot savants, their arrival meant to ground the creative power of Quaoar's Divine Chaos on Earth following Pluto's orgy of decadence, materialism, and decay.[19] Indigos take no prisoners; they have no sales pitch or a plan: they do what they do and it's up to everyone else to figure it out and pick up the pieces. It's not their fault that no one tried to get away with this sort of stuff before. They are conditioned only by what they absolutely know, not by authority or prevailing attitudes. Autism is precisely that: as totally open to *anything* as it is totally closed to *anything else*.

Makemake

Makemake, discovered by Brown's Caltech team on March 31, 2005, is about two-thirds the size of Pluto. In the populous Kuiper Belt, all the dwarf planets collect tiny satellites except Makemake, so there is no yardstick by which to estimate its mass. A minus-470-degree-Fahrenheit iceball of methane, ethane, and nitrogen, it is presently (2015) just about at its aphelion of 53 AU with an orbital period of 309.88 Earth-years.

Named after the creator god of the Rapanui people of Easter Island, Makemake operates astrologically in concert with Haumea and is associated with utilitarian derivations of nature and fertility, both benefic and malefic. The latter includes biotechnology so, at its best and most prophetic, Makemake is the planet guiding us not only into but through the afflictions of Monsanto, Syngenta, and Dow Chemical, ideally into vital permacultures beyond. Even corporate abominations are sacred in origin and have spiritual freedom to create their own universes. Only the god responsible for incarnating these malignant forms can dissolve and transform them, for the universe ultimately needs all its frequencies. Like Quaoar, Makemake oversees the entry of indigo children.

Eris

Orbiting the Sun erratically from 38 to 98 AUs every 557 years, Eris, at 1,450 miles' diameter, is the largest dwarf planet and the ninth-largest body in the Solar System after Mars, Ganymede, and Mercury (six, seven, and eight, respectively). It and its moon are also presently the most distant known objects in the Solar System. When Eris was discovered by Brown and crew on January 5, 2005, he almost named the gray methane world Xena after the fictional warrior queen and then almost named it Persephone, but that tragic goddess had been dibsed by an asteroid. Instead, the new orb was identified as the goddess of strife and discord. Dysnomia, lawlessness, is the Eridian moon.

Persephone is astrologically correct too. The daughter of Demeter, protector of agriculture and fertility, she was snatched by Pluto, ruler of the Underworld, through a crack in the Earth's mantle during an earthquake. In swallowing four of her abductor's pomegranate seeds, she obligated herself to spend four months of the year in his kingdom, Queen of Homeric Hades, overseer of the Gaian winter.

As the dwarf-planetary sublimation of Mars, Eris is "very strong, volatile, and vicious"; she hates to be left out of anything, especially good old fracases and brouhahas. Acting as psycho as the situation allows, she is the last god to leave the battlefield, as she feeds off the mayhem and madness of destruction while having the best time of all.[20] No wonder she was almost named Xena.

As far out as she travels, Eris opens the next gateway or Source Dimension: the quest of the individual soul as human spirit, for its true nature. In the conjunction of Pluto and Eris (December 16, 1756, with an extended footprint from 1755 to 1758), it deposited an early clue to Earth's new direction—not so much the arrant French and Indian War or Russian invasion of Berlin as something camouflaged beneath the return of Halley's comet and the first lighthouse and lightning rod. Eris/Persephone is the aspect of the soul that we now most need, the cipher to open materialism's gate into *what matter actually is.* Ultimately, as the feminine side of Mars, she will teach us how to neutralize his martial aspects in ourselves, maybe even end war forever.

Humanity must break with its cult of materialism before it can turn swords into ploughshares, because to flip weapons of "mass destruction" into windmills, solar panels, teleportation devices, cold fusion, and remote-viewing machines requires nothing less than *collective human telekinesis.* As things presently stand, humanity has substantially abandoned all internal sources of knowledge and guidance, especially those that come from love, clairsentience, and intuition, in favor of simply blasting its way through matter with more ingenious machineries devised by mathematicians and engineers under the sponsorship of more and more impenetrable political systems run by robber barons and their propagandists and lobbyists. The leaders of our species no longer believe in solutions derived from the heart, the third eye, and the aura—or that these are not only more effective ultimately than machines but are the latency from which machines originated, before they began making themselves. When we give over our greater powers to a constabulary of mere objects, we forfeit our Kuiper Belt heritage and destiny. Only a force leveraging the system from as far out as Eris can get deep enough into the human unconscious to find the buried keys and annul the present quarantine.

As unbearably combative and entitled as Eris gets, using every ounce of her appanage to dethrone Pluto, she ultimately brings the cosmos's decisive gift to Earth, though I can't imagine what that is. "In the universe, there are things that are known, and things that are unknown, and in between, there are doors."[21] Eris is a door.

Sedna

Meandering between 76.36 and 937 AUs in a long, highly elongated orbit of 11,400 Earth-years and nearly as red as Mars, Sedna is momentarily closer to the Sun than Eris but, on its oblong course, will pass its neighbor sentry in 2114, at which point Sedna will be the farthest presently known "planet" orbiting Sol. Sedna was discovered by Mike Brown and two associates on November 14, 2003. About 620 miles in diameter, it is the fifth-largest dwarf (after Eris, Pluto, Makemake, and Haumea) as well as a scattered-extended object, meaning that it has been positioned ellipsoidally not by Neptune or Uranus but some unknown celestial factor: perhaps a passing star hundreds of millions of years ago or the hidden perturber of Neptune and Pluto. The ice giants merely refine Sedna's orbit.

Even as I finish this book, new dwarf worlds are being discovered in the cometary cloud, bodies large enough and of suitable composition for their own gravity to mold them spherically—the baseline qualification for planethood. These Oort orbs are so faint—close to pure stealth bodies given how little sunlight they reflect—astronomers now believe that naturally camouflaged worlds as large as Mars and Mercury (and perhaps Earth too) likely abound at the edge of the Solar System. We may yet find Planet X, in single or multiple form.

In March 2014, Trujillo announced 2012 VP113; it is 280 miles wide with an apparent orbit between 80 and 452 AUs. Astrologies for it and other, undiscovered outer worlds are forthcoming; all we can know for sure—and this is a good thing, in fact the only reason that Earth has a chance—is that we don't yet get the full picture: *our* full picture or *its* full picture or how the two of them coalesce. Each new planet, even undiagnosed, is an inkling, an indication of a radically evolving human situation. I would guess that one shared aspect of these dwarfs, considering their bare literal reality, is to bring us into contact with the rest of the universe, in all its dimensions, frequencies, intelligences, and planes. They are our future keys, telegraphically and in every telepathic and telekinetic sense too.

Just as 937 AU is a big nut to turn, Sedna itself wheels a cosmology well beyond our reckoning. Yet there are hints. As a member of the inner Oort Cloud, it seals the zone between the Oort itself and worlds beyond, esoterically. Sedna is the "local" guardian of fractals as well as the crossroads between conflict and emptiness, between all arguments within and among events in the Solar System and the non-argument beyond the Solar System in which the System is premised. As a meta-planet, Sedna is the link between the heart chakra of the inner worlds and the twelfth chakra of outer trans-Neptunian objects. The warmest it gets on Sedna is minus 400 degrees Fahrenheit.

Sedna is the Inuit goddess of the deep sea and marine animals and the queen of the Eskimo Underworld wherein she is known alternately as Big Bad Woman and Skeleton Woman. The designants derive from her primordial confrontation with her father, who became so enraged with her that he threw her off a cliff into the ocean where fish ate out her flesh and eyes.

In another version, Sedna marries Raven and is domiciled by him on a desolate island. When her father comes to visit her, he grows angry at her shabby treatment and intercedes and tries to take her back, but as they are fleeing in his kayak, Raven causes such a terrible storm that Sedna's father, to save his own life, throws his daughter overboard. She clings to the kayak with her hands, so he cuts them off with his knife. Falling into the sea, they become fishes, whales, and seals. Handless, she sinks to the bottom.

In both versions, after lying in the mud for aeons, she is mistakenly pulled up by a fisherman who intrudes into her enchanted inlet. As he hooks her rib cage, the rest of her follows like puppet strings, and she is scooped into his net, a cracked skeleton still bearing long tangled hair. Crustaceans have formed on her teeth, and coral creatures live in her eye sockets. Terrified by what he has raised to the surface, the fisherman tries to fling it back in the ocean, but the bones, caught in his net, won't come loose and are dragged along behind him as he paddles frantically and then races in zigzags across the snowpack. She ends up in his igloo.

Gradually overcoming his fear and loathing, he feels compassion for this being and begins to pay attention to her and her plight. He wraps furs around her skeleton to warm it and then painstakingly restores her bones in their proper order from the toes up. He combs out her hair.

Then he falls into a deep sleep during which he cries a single tear. Crawling passionately toward this droplet, Sedna laps it up. There is so much liquid in the tear that she keeps drinking and drinking until she has quenched her long thirst. After this quaff she looks into the fisherman's chest, finds his heart and extracts it, and, using it as a drum, beats out a rhythm to which she supplies words: "Flesh, Flesh, Flesh! Flesh, Flesh, Flesh!"

The aria restores tissues to her bones, eyes to her sockets, hands to her radii. As she continues to beat and chant, she grows breasts and a vagina. Then she sings the clothes off the fisherman and reaches back into his chest, where she returns the drum so that he has a beating heart. She presses her full womanly length against him, falls asleep, and they awake in the morning tightly entwined.[22]

In 2009 I met a fairly convincing alien in a dream who told me, after I asked him whence: "Orion. Actually one of the stars in what you scope as Orion."

I stared at him for a long time before deciding he was real. Then I set before him the crucial riddle: "How did you get here? You had to beat the speed of light, right?"

"Light is not," he beamed, "my measure. We have something called"—and I forget the acronym, maybe BLIP or FLIB—"it transfers information, and then I come along. First I put my data into your system; then I instantaneously appear."

"Is it like remote viewing," I wondered, "only remote super-positioning?"

"It is," he said. "But what I actually do is more complicated and stranger than you are assuming. The trick is to get from your kind of physics to mine. Also I had to discover this place in the atlas of worlds in order even to know it existed; then I sent my coordinates ahead."

He didn't say that or even give that speech, but it is a fair rendering.

As I lost touch with his semblance, I felt the hefty premise of his great science against our world's colossal blundering. I did not get to ask him why he had undertaken his trip or what he thought would be our outcome.

The next night, though, I woke from a dream with his ambiance inside me and the following message about the stations of our System:

Mercury: Information portal and entry strip for transdimensional travel

Venus: Electromagnetic node and operational symbol for the post-petroleum era of Gaian civilization

Earth: Alchemy then, alchemy now, alchemy forever

Mars: The ancient past of our Solar System with its pre-Egyptian, pre-Lemurian civilizations

Jupiter and its moons: The life forms, consciousness, and buddhas of the future

Saturn and its moons: The unknown—beyond all life forms and imagined life forms, beyond all worlds and imaginable worlds

Uranus and its moons: The beginning of the absolute next thing, the eternal return of *this* thing transmogrified and altered by moving one critical atom at its core

Neptune and its moons: Peace, peace forever, peace at last

Pluto and Charon, the Kuiper Belt, the Oort Cloud: Communication and commerce with the rest of the universe

Notes

[1] Sidney Carroll and Robert Rossen (from a novel by Walter Tevis), *The Hustler*, directed by Robert Rossen, 1961.

[2] Johann Galle, quoted in William Graves Hoyt, *Planets X and Pluto* (Tucson: University of Arizona Press, 1980), 53.

[3] Sir John Herschel, quoted in Richard Berendzen, Richard Hart, and Daniel Seeley, *Man Discovers the Galaxies* (New York: Science History Publications, 1976), 184.

[4] Henry Norris Russell, quoted in Hoyt, *Planets X and Pluto*, 231–32.

[5] Hoyt, *Planets X and Pluto*, 161–62.

[6] Ibid., 162.

[7] Percival Lowell, quoted in ibid., 141.

8. Vesto Slipher, quoted in ibid., 179.

9. Clyde Tombaugh, quoted in ibid., 188.

10. Dale Cruikshank, Carl Pilcher, and David Morrison, quoted in ibid., 245.

11. Fritz Brunhübner, "Pluto" (1934), reprinted in *Io,* no. 14 (Earth Geography Booklet, no. 3: Imago Mundi), edited by Richard Grossinger (Plainfield, Vermont, 1972), 281–90.

12. Oskar Seyffert, *A Dictionary of Classical Antiques* (New York: Meridian Books, 1956), 263.

13. Fritz Brunhübner, "Pluto" (1934), reprinted in *Io,* no. 14 (Earth Geography Booklet, no. 3: Imago Mundi), edited by Richard Grossinger (Plainfield, Vermont, 1972), 281–90.

14. Ibid.

15. William (a.k.a. Ellias) Lonsdale, unpublished talk, Goddard College, Plainfield, Vermont, 1977.

16. James Hillman, *The Dream and the Underworld* (New York: Harper and Row, 1979), 28.

17. Nick Anthony Fiorenza, "The Astronomy and Astrology of Ixion." www .lunarplanner.com.

18. Denis Kutalyov, "Angel or Demon?," http://astrologic.ru/english/Varuna .htm.

19. Richard Brown, "Karmic Astrology: "Quaoar by Sign, A Generational Indicator," http://astrology.richardbrown.com/NewPlanets/qgen.shtml.

20. Midlands School of Astrology, "Eris: Goddess of Strife, Stimulation & Rivalry," www.midlandsschoolofastrology.co.uk/eris_goddess_of_strife_ and_stimulation.html, and "The Astrology of Eris," www.historicalastrology.com/the-astrology-of-eris/.

21. This quotation has been attributed to both William Blake and Aldous Huxley, but the closest form to the version I use comes from Ray Manzarek, cofounder (with Jim Morrison) of The Doors, who said it in response to an interview question by a *Newsweek* reporter for a profile on the group in 1967.

22. The Inuit tale of Sedna is from a number of sources, including Clarissa Pinkola Estés, *Women Who Run With the Wolves: Stories of the Wild Women Archetype* (New York: Ballantine Books, 1992), 139.

3

New Horizon... for a Lost Horizon

RICHARD C. HOAGLAND

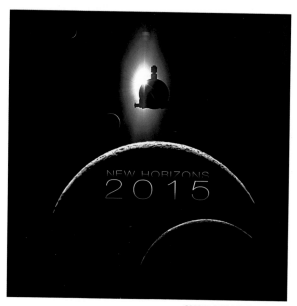

Figure 1. The New Horizon Mission is humankind's first modern visit to Pluto—the last "classic" planet of the solar system. In this Mission artwork, the spacecraft is depicted as "eclipsing" the distant sun . . . a "coded" clue to the HD/Torsion Physics the New Horizons spacecraft will be quietly measuring as it flies through the unique Plutonian system of a true "double planet" . . . orbited by (at least) four "harmonic" moons. . . .

When Richard Grossinger initially asked me to consider contributing to this unusual volume, in anticipation of NASA's first unmanned mission to Pluto (called *New Horizons*—finally flying by the controversial "ninth planet" this summer), I remember that my first response went something like—

"But how does Pluto—a tiny, demoted 'ball of ice' (no longer even considered officially 'a planet')—fit into our decades-long Enterprise investigations into the profound possibility, growing more certain with every new 'leaked' NASA ritual and image, that 'an ancient, highly-advanced ET civilization' a *long* time ago systematically *remodeled* the entire solar system—with *us* as the eventual result!?"

This chapter is the surprising answer to Richard's invitation.

With appropriate homage to James Hilton and Frank Capra,[1] our "lost horizon" and the "new horizon" that is about to overtake it encompass some remarkable and surprising parallels to their 1930s classics:

In the 1937 film adaptation, the idyllic valley of perpetual peace, youth, and perfection—Shangri-La[2]—is depicted as a timeless mountain valley refuge, deliberately placed high amid the "impenetrable Himalayas," to attempt, if all else fails, preservation of its priceless archive of "ancient human knowledge" against the relentless tide of another global war.

That fictional "Lost Horizon" scenario is, in fact, an eerie echo of the results of our own, now decades-long, Enterprise Mission investigations regarding the growing probability of a "totally *remodeled,* ancient solar system" whose all but godlike "ET knowledge and accumulated wisdom" may have been preserved—across millions of years—in a series of "dedicated archives" cached across that ancient solar system.

Including the one *most likely* to have survived—at Pluto.

The Enterprise Mission, across more than thirty years, has published[3] a variety of increasingly compelling evidence supporting these radical ideas—that this entire solar system, somehow in the dim and distant past, was deliberately "rearranged" to ultimately support the development of human intelligence ... and eventually human civilization itself.

First outlined in 1987 *(The Monuments of Mars: A City on the Edge of Forever),*[4] the academic science behind this far-reaching Enterprise hypothesis seems to have finally, quietly, caught up.

Recent peer-reviewed astronomical observations and analysis—stemming from NASA's revolutionary Kepler space telescope "exo-planetary" observations of the numbers and types of "super-earth planets" found orbiting even "nearby, sun-like stars"[5]—have reached a startling "mainstream" conclusion:

> It seems clear that our [own] Solar System—which contains no planet interior to Mercury's $P = 88$ day orbit—*did not participate* in a major if not the *dominant* mode of planet formation *in the Galaxy* ...

Super-earths [planets in the range of five to ten times the Earth's mass] are *not* anomalous; they are the rule *that our Solar System breaks*. In a sense, *the burden of explaining planetary system architectures rests more heavily on the Solar System [now] than on the rest of the Galaxy's planet population at large* [emphasis added].[6]

Britain's *Daily Mail,* reporting on this same revolutionary astronomical development, was more direct:[7]

In the last 20 years, reports *Nature*[8] . . . it may even be that our solar system is *fairly unique* when compared to other [Galactic . . .] planetary systems [emphasis added].

This, of course, is blatant scientific heresy, standing against over half a *millennium* (since Copernicus) of emphatic and repeated scientific assertions that the Earth (and its parent solar system) must be "relentlessly average" (the most often cited example of the so-called "Copernican," or Mediocrity Principle)[9].

The fact that NASA has now discovered (and publicly admitted!) that the solar system, for unknown reasons, does *not* follow science's "prime mediocrity directive"—and is now being viewed by more and more astronomers as anything but "average"—only foreshadows a much larger revolution to come.

This dawning paradigm shift—that there is something uniquely "wrong" (or, right!) about *this* solar system, compared to almost all the thousands of others currently known—after two decades of "exoplanet" observations, in our opinion, takes our thirty-year-old Enterprise idea of a possible Type II Civilization "re-arrangement" of *this* early solar system from being merely "a far-out possibility" to the level of another *scientifically-testable* "model."

A hypothesis that could turn out to be "*the* explanation" for the series of extraordinary solar system and exo-solar system discoveries that NASA (and the other international space agencies) have quietly . . . secretly . . . been finding and then rigidly *suppressing*—for more than fifty years.

There are verifiable remains, scattered all around us in the solar system, of a stunning Type II Kardaschevian civilization;[10] hundreds of millions of years after radically rearranging the entire "original" size and spacing of the planets and moons orbiting the Sun (as in "Bode's Law," et al.), this civilization was *murdered* through a series of cataclysmic, literally "world-shattering" *inter-planetary* wars. From this destruction, a few priceless fragments—across countless ensuing millennia—have been carefully handed down through time by each "high civilization" that successively rose up, like the original Egyptian phoenix, culminating now (on Earth) with *us*.

It is these *sacred*, deeply-sequestered "ancient ET archives"—as we have argued elsewhere[11]—that are now revealed, from our continuing analysis, to have formed the underlying "hidden guidance" behind NASA's *entire* first fifty years of pioneering manned and unmanned exploration of the solar system!

Each successive NASA mission over that "Golden Age" of intensive exploration was selected, we have shown, *not* primarily for its "potential to advance mainstream 'geological planetary models,'" as NASA has always flatly claimed, but as part of a larger reconnaissance of a secret . . . *sacred* . . . ancient list of long-abandoned *ET spacecraft and planetary ruins,* to be scrutinized by a series of key NASA "cover" missions. These missions have revealed startling close-ups of miles-wide, *war-shattered*, still outgassing fragments of that once godlike Type II Civilization's ancient space fleet—indistinguishable, to pre-space age generations of Earth-bound astronomers, from all the other interplanetary "asteroids and comets."

Our research has now revealed that this stunning, *new* solar system reality first became known to the U.S. Department of Defense under the Eisenhower Administration in the 1950s—in part via a pioneering, top secret JPL (U.S. Army) effort at a first unmanned circumlunar reconnaissance mission, "Project Red Socks." We believe the shocking results of this clandestine mission formed the *real* reason behind NASA's sudden public emergence after Sputnik, and the rapid congressional authorization, only one year later in 1958, of NASA as the loudly-proclaimed, lead "*civilian* space agency of the United States of America."

This was the perfect cover—in the 1950s world of perpetual Soviet pursuit of *any and all* technological supremacy over the West—for NASA's *real,* long-term covert mission:

To secretly ascertain . . . from NASA's inception . . . the full extent and (potential) *military* threats (or benefits) of these long-abandoned, ancient ET *derelicts,* still endlessly "falling around the sun," as well as those ancient surface installations still partially preserved on various planets and moons

(Cydonia et al.); the surviving riches of an entire, astonishing Type II Civilization in our own backyard—whose extraordinary legacy and scientific potential was only fully accepted (even within NASA) when Apollo astronauts fulfilled their *real* Kennedy Mission and clandestinely returned, beginning in 1969, unquestionable *intelligently-designed and manufactured ET artifacts* to Earth—from the Moon.

A "program . . . within a program . . . within a program"—all carried out in plain sight—under the rigidly repeated NASA mantra: "the Agency is *only* pursuing a *civilian* lunar and planetary science program."

"*Stargate: SG-1,*"[12] anyone?

CREDIT: NASA/THE NEW HORIZONS MISSION

Figure 2. Artist depiction of a "cold, sunny day on Pluto" with Pluto's one large moon, Charon, looming just ~12,000 miles away through an ultra-thin, ultra-frigid (~-380 –degree F.) atmosphere of nitrogen, methane, and carbon monoxide, outgassed recently from Pluto's perpetually frozen landscape.

Pluto—formerly the ninth planet—was confirmed as an official new member of the solar system on a cold February night in 1930, on a mountain plateau in northern Arizona, by a young twenty-three-year-old former Kansan amateur astronomer named Clyde Tombaugh[13] (that little "dust up"[14] over at the International Astronomical Union as to whether Pluto is a "planet" notwithstanding; we'll get to a possible real reason for the very curiously timed, and totally unnecessary controversy, later).

Tombaugh, a self-confident but quiet young farmer, financially prevented by crop failure from entering college to pursue a career in astronomy, had mailed his own astronomical drawings—particularly of Mars and Jupiter—to the Lowell Observatory for "review" anyway in 1928. To his surprise, as a result of this Midwestern initiative (combined with genuine talent), Tombaugh was immediately hired by the Observatory's director—Vesto Slipher—to carry out a "mission."

Tombaugh's "mission impossible": to promptly "resume a systematic photographic search for a possible 'ninth planet of the solar system,'" initially started by Lowell's founding director, Percival Lowell himself, in 1905.

The original Lowell Observatory search program[15] had been based on telescopic measurements Lowell painstakingly carried out of the orbital position of the outer "gas-giant planets," Uranus and Neptune, as compared to their *calculated* ephemeris positions—after Uranus's (and all the other planets') slight, competing gravitational attraction on Neptune had been carefully factored in.

The two didn't match.

Which meant, according to standard (Newtonian) interpretation, there *had* to be a (still unseen) "new" planet out there . . . somewhere—a "gravitational perturber" tugging more strongly on Uranus and Neptune than on the rest of the known solar system—*a genuine "ninth planet."* Based on his observed "Neptune and Uranian discrepancies," Lowell later wrote that he immediately began calculating where on the celestial sphere an unseen planet "of about six Earth masses"—which he called Planet X—might lie.

Twenty-four years later, Clyde Tombaugh—the enthusiastic amateur-turned-professional at Lowell—would finally, quietly, vindicate Lowell's amazingly predictive calculations.[16]

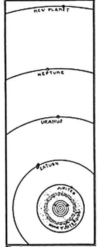

The newspaper clipping reads:

179

The Sun's New Trans-Neptunian Planet

Lowell Observatory Confirms Its Founder's Prediction

By Dr. V. M. Slipher
Director, Lowell Observatory

THE Lowell Observatory has made the discovery of a celestial body whose rate of motion and path among the stars indicates that it is a new member of the sun's family of planets out beyond Neptune.

Twenty-five years ago, Dr. Percival Lowell, director and founder of the Observatory at Flagstaff, Ariz., began a mathematical investigation for a planet beyond Neptune, based upon certain unaccounted for motions of the planet Uranus. The problem of locating such a body in the heavens was a very difficult one, and involved an enormous amount of intricate computations. In 1914 he announced in an extensive memoir as a result of his mathematical work, the position of the predicted body.

The search of the sky directed by Dr. Lowell's theoretical investigation was begun by photography in 1905 and has been continued with interruptions to the present time. Use has been made of the best available instruments covering that band around the sky in which the planets travel. Early in 1929 the new Lawrence Lowell telescope, a special instrument for the research, was put in operation.

Some weeks ago Mr. C. W. Tombaugh found an object on his plates with this telescope, which has since been followed carefully also with the Lowell forty-two inch reflector, by C. O. Lampland. It has been observed visually with the large refractor by the various members of the Lowell Observatory staff.

The object is now (March 13) about twelve seconds west from delta Geminorum, westward motion about two seconds per day. The position of the new object substantially agrees with Lowell's prediction of its position and distance.

The discovery of a new planet beyond Neptune by Lowell Observatory is one of astronomy's outstanding discoveries. The year 1930 will be remembered for the detection of the ninth planet.

The diagram shows the relative distance of the new planet from the sun. From the sun outward, the unlabeled planets are: Mercury, Venus, Earth, Mars.

A cranberry placed 3¼ miles from a globe two feet in diameter make a model of the new planet in relation to the sun.

By Clyde W. Tombaugh
Discoverer of New Planet

IN searching for the new planet I was carrying out a systematically arranged program and was fortunate in being assigned to this work with the splendid new Lowell photographic telescope. I was determined to examine the ecliptic thoroughly and carefully all the way around the sky. In the course of several months of arduous research I had been ever expecting to find the new planet predicted by Prof. Lowell.

Examination of plate after plate failed to reveal it, but many asteroids and variable stars were found. I had figured out just how the object sought for should appear. The ecliptic survey was nearly half completed when one day I found an object on my plates that fulfilled expectations. Almost instantly I felt that it was the one looked for, and, of course, felt greatly elated. I was reminded of my sister's class prophecy back in high school days.

The work on the planet, however, is far from finished. Now that it is found, the elements of its orbit, and much else concerning it, must be learned, so doubtless it will be a much observed object. I am not a mathematician, and so the work on the planet is being carried on largely by the senior members of the observatory staff.

I was born February 4, 1906, near Streator, Ill., the son of Muron and Della Tombaugh; received my elementary education in the rural schools and attended Streator High School for two years. The family moved to Kansas in 1922, where I assisted my father in raising wheat. I was graduated from Burdette High School in 1925.

Since then my summer seasons were given to farming, my winters to constructing reflecting telescopes. During

CREDIT: LOWELL OBSERVATORY

Figure 3. Original Lowell Observatory press announcement of Clyde Tombaugh's 1930's confirmation of Percival Lowell's 1905 "ninth planet prediction."

The name *Pluto* was announced three months after Tombaugh's historic discovery photographs were verified, in May, 1930; the story of the time was that "an eleven-year-old girl in England, Venetia Burney, connected to Oxford University through her grandfather, a retired Oxford librarian," was able to get a "cable" (telegram) to Tombaugh in Flagstaff just days after the discovery announcement," urging him to "name his new discovery 'Pluto' … for mythological reasons."[17]

As the story continues, Tombaugh liked it … and "the rest is history"; the added advantage being that the first two letters—PL—"in code," also commemorated Percival Lowell.

In an important reinforcement of "the symbolic overtones" to Pluto's discovery confirmation story, evidence has emerged that purported "33rd Degree Freemason," Walt Disney[18]—creator of the most popular animated animal and cartoon characters of all time—having launching a planned new "canine character" (in the fall of 1930), within months of Pluto's discovery and naming suddenly changed the name of his new dog (from "Rover" in 1930 ... to "Pluto" in 1931)[19]—to take "deliberate advantage of the public's obvious, overwhelming interest in the new planet."

Thus, Pluto—soon, the new "top dog" of pop film culture around the world—was also subliminally, inextricably "imprinted" in the public mind (through the Disney "media machine") as that new "'doggy' planet" ... a celestial "canine connection" that was an obvious echo of a much more ancient, much more powerful celestial canine meme: the original Egyptian worship of the "dog star" (Isis) herself: Sirius.

"Symbolic memes" and NASA's repeated "inexplicable obsession with Egyptian mythology" aside, unfortunately for Pluto's long-term historic legacy, physical doubts regarding the new planet's fundamental nature soon began emerging, despite Tombaugh's initial, apparent triumphant vindication of Lowell's decades-old "ninth planet orbit calculations."

Independent telescopic observations coupled with rapidly developing astronomical technology soon made it clear that the "Lowell/Tombaugh" Pluto was *far too small* (thus lacking the crucial gravitational field from not possessing enough mass) to have *possibly* created the "celestial mechanics errors" Lowell had reported in his initial analysis of Uranus's continued "anomalous orbital motion" (again, after the rest of the solar system's known gravitational influence, particularly that of Neptune, was carefully subtracted). These celestial mechanics anomalies, ostensibly, were the entire observational basis for Lowell initiating his own 1905 "high-profile, formal launch of a Lowell Observatory search for a ninth planet!"[20]

So—

If Lowell's *measured* outer planet discrepancies turned out to be "so wrong," how did he still get the answer, regarding the *location* of Pluto decades later, so *right!?;* how did Lowell successfully *predict* the celestial position of a small (ultimately, as it turned out, 1,400-mile-wide) "snowball," at least four billion miles away, fifteen years ahead in time? An object which could not *possibly* (from a wide variety of modern astronomical measurements and subsequent analyses) have generated the "anomalous gravitational effects" on Uranus and Neptune that Lowell reportedly initially *measured?;* that he specifically (according to his notes) incorporated into this bafflingly "correct" (even if now *proven* mathematically wrong) "discovery position" for Pluto?

Was there, in fact, more to Lowell's "impossibly" successful prediction than we've been told?

Current academic astronomy casually dismisses Lowell's fifteen-year prediction of Pluto's distant location as "just a lucky coincidence":

> The question arises … why is there an actual planet moving in an orbit which is so uncannily like the one predicted? . . . There seems no escape from the conclusion that this is *a matter of chance.* That so close a set of *chance coincidences* should occur is almost incredible, but the evidence employed by Brown permits of no other conclusion [emphases added].
>
> —RICHARD RUSSELL, Princeton astronomer, 1930[21, 22]

This "religious" academic approach (it *has* to be "chance") obscures the *real* issues nagging Lowell's uncanny 1915 Planet X calculations for over a hundred years: What if his correct prediction wasn't "just an incredibly lucky accident?"; what if there was a *third* explanation?

If, as has been shown, "the laws of gravity" *totally rule out* the current "Pluto" as being, at all, physically responsible in any way for Lowell's published Uranian/Neptunian discrepancies, then the only rational (if politically unacceptable) *third* remaining explanation for Lowell's astonishing "success," would be if he (Lowell) literally *made up the "Neptunian/Uranian anomalies!*

So—

He could quietly "insert" a new, *real* planet (read "Pluto") in the outer solar system … *in the correct orbital* location. . . ultimately to be "found" by anyone following Lowell's carefully planned (but—in this theory—*deliberately* misleading) "Neptunian/Uranian discrepancies"—which was "necessary" (in this theory) *if* Lowell was to successfully hide his *real* "off-world *source"* for such an astonishing, bona-fide solar system addition!

An *authentic* ET archive.

In fact, it is quite reasonable to propose that Lowell—as a very "well-connected" Boston "Brahmin"[23] who had personally lived in the Orient for many years, who had thoroughly cultivated a wide variety of local contacts in the secular and temple communities (writing several best-selling volumes in the process on a variety of Asian subjects, which are now credited with "opening up early twentieth-century American appreciation/awareness of the looming importance of the Far East")—may also have gained, as "a well-known American amateur astronomer . . . *with money*," privileged access to an "Asian version" of that proposed "secret . . . ancient . . . *sacred* . . . ET solar system list!"

A list on which a future "Pluto" (and its critically important, *current* celestial position —an accurate "planetary ephemeris") was, in the model, quietly preserved—plainly identified, perchance, as "housing a rare, surviving solar system Ancient Archive . . . connected with a, now-extinct, ancient *civilization* once spanning the entire solar system . . . including, ruins still visible on Mars."

The existence of such a mind-boggling list . . . or Lowell's proposed "secret access," notwithstanding, what is historically indisputable is that Percival Lowell suddenly, dramatically, *returned* to Boston in 1893—after ten years of continuously living and writing in the Far East—and *immediately* began refocusing his "entire wealth and managerial genius" on "The Heavens."

Along the way, he created (in less than a *year*) the first university-grade (but independently private) American "Lowell Observatory"—located, for the first time, *thousands* of miles away from the East Coast centers of "academic science," especially Boston—in the "Great American Southwest" amid the high, "astronomically perfect" mountains of northern Arizona.

And, unlike the demonstrated "polymath" he'd been throughout his entire previous career(s)—"shrewd investor . . . skilled photographer and scientific artist . . . inspired public speaker . . . and immensely popular adventure-travel writer"—Percival Lowell, on abruptly, radically, changing *his entire life* from all that he'd been previously, for some extraordinary personal reason decided to devote his last two decades to one thing and one thing *only*—

Mars.

Figure 4. Percival Lowell eloquently compares the "saffron Martian deserts" he is seeing through his specifically-designed 24-inch telescope, when he points it at Mars, to those same "saffron hues" surrounding his Southwest mountaintop observatory on "Mars Hill," looking down on Flagstaff, Arizona.

Lowell began a single-minded visual and photographic search, painstakingly pursued over *decades* . . . for scientific evidence of "Life on Mars"; a search morphing (in his last decade) into a parallel (but, as you will see, equally "obsessive," if potentially *connected*) search for "the location of a [then uncertain] ninth solar system planet."

This raises the crucial question: What (if anything) was Lowell *physically* shown in Japan to make him rush home and begin (literally) *a whole new life*—that gave him such conviction regarding what was *really* waiting for humanity on Mars . . . if not beyond?

And, did Lowell's unshakable scientific certitude on these *two* issues—Mars and Pluto—as our years of Enterprise analysis now concludes, stem from the *same*, "ancient (ET) 'sacred' solar system sources" as NASA's ... only *decades* earlier!?; did Lowell, in fact, *deliberately* "leak" these two new, major clues about an inhabited ancient solar system from this "ancient, secret list"?

Perhaps as part of a larger, clandestine *political* effort in the early twentieth century—aimed, worldwide, by a whole group of "renegade intellectuals"—at *forcing* an eventual "democratization of *all* ancient solar system knowledge and information" held secretly, for countless previous millennia, by "the elites"? The *first* "Disclosure Project?"

Was *this* why there was such an overwhelmingly negative, organized "academic backlash" against Lowell's late-nineteenth-century/early-twentieth-century Mars observations, and certainly his relentless pursuit of Martian "life?"

These controversies (if possible) only escalated around Lowell's sudden decision to also search for Planet X. And they ultimately led, in 2006, to the very planet he had successfully, mathematically predicted decades before being deliberately, *politically* "demoted" . . . just as it was finally about to be *robotically* explored.

In this most recent "Lowell humiliation"—the highly public political "demotion" of Pluto—was Percival Lowell being "publicly punished" *once again* . . . because, as an insider, he "broke the Ultimate Rule?" Was *this* the real reason why Pluto—suddenly, after decades as the solar system's official "ninth planet"—was *institutionally* "dethroned" to being just *one of* "thousands of so-called "dwarf-planets" orbiting just *beyond* Neptune's 30 AU orbit in what is called the Kuiper Belt?

Figure 5. The so-called "Kuiper Belt" is a vast, donut-shaped ring consisting of trillions of (newly-discovered in 1992) "icy-asteroids" of a range of sizes from "planetary" all the way down to "dust" —starting just beyond Neptune's orbit (30 AU) and extending out to about 50 AU. Pluto, because of its orbital positioning and the IAU "decision, was controversially "dethroned" (in 2006) from being the Ninth Planet to being "just another member of the Kuiper Belt."

CREDIT: NASA

Was this sudden demotion of Pluto into being "just another Kuiper Belt Object" (by, in fact, just *4 percent* of voting IAU members!) "coincidentally" just *weeks* after a NASA mission finally got underway for the first close-up look in modern history at the last "classical" planet—in fact, just a thin, if not obvious attempt to begin reducing public *interest* in Pluto well before *New Horizons* finally got there? Again, this was perhaps part of a larger, continuing twentieth/twenty-first-century academic effort to totally discredit *everything* that Percival Lowell ever discovered, because Lowell's "source"—like NASA's—was ultimately a forbidden *"off-world archive?"* Was this whole modern kerfuffle around Pluto, ultimately, simply because the Powers That Be felt Lowell needed to be made "a *continuing* Example"[24] of what happens (even to "Brahmins") who *reveal?*

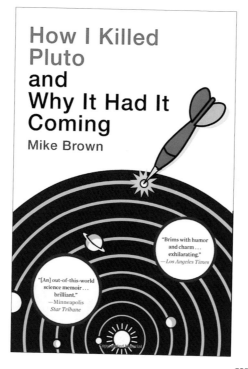

CREDIT: MICHAEL BROWN

Figure 6. Cover of CalTech astronomer Michael Brown's "in your face" defense of the "dethronement of Pluto."

This curious (and totally disturbing!) pattern of "continuing Lowell's punishment" seems to extend even beyond Lowell's personal scientific achievements; the seminal "extra-galactic nebular observations" of Vesto

Slipher[25] (Lowell's personally appointed successor as director—and the same guy who would later hire Clyde Tombaugh to "resume Lowell's ninth planet search") have also been quietly "back benched" in history, by powerful "academic interests" still politically opposed to *Lowell*.

Slipher was the early twentieth century's preeminent "astronomical spectroscopist"—the term for an astronomer of the time who specialized in studying the chemical composition of and physical conditions surrounding large, nearby objects in space (such as the solar system's other moons and planets) by carefully analyzing how reflected sunlight is changed (imprinted) when it interacts with their various surfaces or atmospheres.

But planets are (relatively) bright objects, providing adequate light for such insensitive, early-twentieth-century "spectroscopic technology."

A key part of Slipher's 1920s commission, after Lowell hired him for the Observatory, focused on extending such solar system "composition and physical information studies" to *galactic objects,* thousands of times dimmer than the nearby planets. These "interstellar spectroscopic objects" included single (and multiple) stars, galactic nebulae, cataclysmic objects (nova and supernova), etc.—and even (potentially) the ultimately mysterious (because they were suspected, even then, of being *extra-galactic*) ghostly "spiral nebulae."

Slipher, in pursuit of these "cutting edge" astronomical observations for Lowell, had to (literally) invent a totally new celestial science (and supporting optical technology)—which would ultimately become the foundation of a radically new twentieth-century scientific discipline called astrophysics.

But it would be George Ellery Hale[26] (founder of the famed Mt. Wilson observatory high above Los Angeles, California), not Slipher, who would ultimately receive "the scientific credit" for this fundamental twentieth-century astronomical revolution because of his innovative spectroscopic *solar* observations.

Another deliberate slight against Lowell's "people"?

Slipher's pioneering technological genius in "galactic spectroscopy," and, as a direct consequence, his serendipitous and startling discovery of "anomalous spiral nebular *redshifts*"—as a singular result of his own major optical and film improvements—would become the observational foundation for the entire, subsequent Big Bang paradigm, of "receding galaxies and expanding universes!"—but, without as much as a *hint* from mainstream academia that *all* this was based on Slipher's *prior*, extra-galactic spectroscopic work . . . at the behest of Lowell!

By giving credit for these shatteringly important early galactic redshift observations (and, ultimately, even a "celebrity telescope!") to Edwin Hubble (also at Mt. Wilson) and not to Vesto Slipher, the political winds of the time, in hindsight, seem pretty obvious: Lowell (and his Observatory) was *never* to be scientifically "acknowledged" again, nor were the folks who chose to pursue their own research under Lowell's direction.

This pattern of "extreme Lowell pettiness," amazingly, seems to be *continuing . . .* apparently, even extending to the current NASA *New Horizons* mission! *Enterprise* has discovered that Clyde Tombaugh's ashes have been quietly included on the solar-system-escaping spacecraft[27]—a fitting beginning to his "post-life forever odyssey" in the *Pharaonic* tradition of "sailing the boat of millions of years . . . to one day 'reach the imperishable stars,'" to, literally, cruise the galaxy forever aboard *New Horizons . . .* beyond Pluto.

Percival Lowell's ashes, pointedly, have *not* been so included.

If those trying to "re-bury Lowell"—and Pluto with him—believe these tactics are working, they need to think again; judging by the overwhelming public firestorm[28] that erupted against Pluto's "academic planetary banishment," if someone thinks just "calling Pluto by a different title," or preventing Lowell from "returning to the stars . . . by leaving him buried in his lapis lazuli mausoleum on Mars Hill" will have *any* diminishing effect on public interest in the *New Horizons* mission—they should talk to Disney.

There was, of course, one other, larger "conspiratorial" possibility to explain this sudden, almost irrational "official war on Pluto" reaching its political climax just after *New Horizons* launched: That the whole, high-profile "political execution" of the ninth planet was actually intended to generate a totally fake *controversy* around "Lowell's Pluto," to deliberately write an "interplanetary soap opera" so that by the time *New Horizons* does fly by this year, a *lot* more bloggers and consumers of social media (if not global TV networks and websites looking for more ratings) will be paying close attention to "the non-planet" and what is found there—if only because of the (soap opera) "controversial planet / controversial astronomer" angle.

Already, the "controversy angle" surrounding Pluto is being pushed in the media as new "focus groups" of scientists are polled as to whether the IAU decision should be reversed—and, surprisingly, the majority have voted that Pluto "should be *reinstated* as a planet."[29]

"Science is nothing . . . if it's not prediction."

Alan Stern—*New Horizons's* principal investigator—has been firmly of the opinion, since 2006, that the IAU "made a terrible mistake in re-classifying Pluto."[30] According to Stern, "The *New Horizons'* Mission . . . will *not* recognize the IAU's planet definition resolution of August 24, 2006" [emphasis added]. Does Stern suspect (or know) the *real* reasons for Pluto's abrupt demotion . . . a reason that is, in fact, connected with *his* own (in our model) "clandestine Pluto mission"—part of the larger, *symbolic* reason on this mission why Tombaugh's remains *could* but Lowell's ashes could not be included on *New Horizons?*

Tombaugh's photographic confirmation of Lowell's Pluto orbit predictions occurred well over fifteen years before I was born; only after several more decades, however, would I finally have the pleasure and distinct honor of discussing "astronomy," one whole afternoon, with Clyde Tombaugh—one of the bona-fide "astronomy icons" from my own growing up and growing scientific curiosity—in person.

Our conversation took place, appropriately, at NASA's Jet Propulsion Laboratory, in southern California.

It was Mars, that time, not Pluto, that had finally brought the two of us together in one auditorium: JPL's "Von Karman, "joined by several hundred fellow reporters and planetary scientists; the occasion had been NASA's historic unmanned *Viking* mission to Mars . . . to become the first "20th Century robotic artifact" to successfully "set an American footpad on the Red Planet," that morning of July 20 in 1976.

My meeting Tombaugh at JPL, just days after the amazing *Viking I* landing, was "icing on the cake" to having the once-in-a-lifetime experience of being physically present to observe, first-hand, modern *Homo sapiens's* Mars Return. . . .

Now, of course, I occasionally ask myself: Where would our afternoon conversation have gone . . . if what I know *now* about Mars, I had known *then?*

That remarkable afternoon I spent with Tombaugh brought to an amazing synchronicity a series of striking personal "nodal points" (as I have often described them) in my own life, all curiously involving "Mars"—that were set in motion for me over a decade before when I went to work for the Springfield Museum of Science, in Springfield, Massachusetts after successfully "passing" a private presentation I was requested to give to the museum director, in applying for the job. My chosen subject: Mars.

But, in perhaps an even more remarkable "Mars coincidence," one of the members of the Springfield Museum Board of Trustees was Roger L.

Putnam—Percival Lowell's own *nephew*, and sole trustee of the Lowell Observatory for more than forty years after Lowell's passing; when the search for
Lowell's "ninth planet" almost died in the early decades of the twentieth century,
after Lowell's untimely death at sixty-one in 1916, it was Putnam who *single-
handedly* revived it over the strenuous legal objections of Lowell's widow—
even raising a crucial ten thousand dollars from Harvard University for design
and manufacture of a new "photographic telescope" that would one day "take
the plates" on which Tombaugh ultimately confirmed "Lowell's" Pluto.

Looking back, it is now clear to me that it was Putnam's persistent (if subtle) personal encouragement behind the scenes, after I met him one night
at a museum reception for a new exhibit, culminating in the not-so-subtle
gift of a signed, personal volume on "Mars and the Lowell Observatory," that
"helped me on the path to Mars" at a seminal moment in my professional
career, if not my life; that gave me the "institutional courage" to undertake
my first "national Mars project" at the Springfield Museum, in the summer
of 1965, focused on "the imminent fly-by of the first NASA spacecraft to
visit the Red Planet—*Mariner 4*.

I called my first official museum exhibit and program "Mars: Infinity to 1965."

Figure 7. Springfield Museum of Science program
book from its "Mars: Infinity to 1965" exhibit, exploring the details and implications of NASA's first
unmanned mission to Mars, Mariner 4. The unique
exhibit was designed and executed by Richard Hoagland as his first "Mars project," when he was the Museum's curator of astronomy and space science.

"Mars: Infinity to 1965" was (for its time)
a "state-of-the-art" multi-media program
exploring "the scientific and cultural implications for Humankind, surrounding the first
NASA unmanned fly-by of the 'fabled' planet
Mars—by *Mariner 4*."

CREDIT: RICHARD C. HOAGLAND

The program included "space paintings" (an exhibit of *original* "Bonestells"),
projected NASA film and TV images, 3-D physical exhibits (including a "hanging solar system," to electrically track the orbit of *Mariner 4* en route to Mars),
and a mock-up "growing plants on Mars experiment" from Union Carbide.

The Mars exhibit I designed (and even physically constructed, in part),
and the Peabody Award-nominated live radio show I co-produced and

presented to New England the night of the fly-by became the first NASA mission I was professionally involved with.

Mariner 4 recorded twenty-two "primitive" television close-ups[31] of the frigid Martian deserts in those few "close-encounter" hours, for playback later . . . close-ups on which it would become shockingly apparent, in the days after *Mariner 4*'s successful fly-by, that Mars—instead of being covered with "an artificial network of Martian-constructed 'Lowellian canals'"— was, instead, covered with millions of ancient *impact craters,* just like the lifeless Moon!

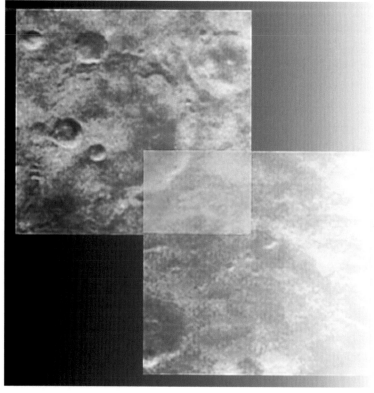

CREDIT: NASA/JPL

Figure 8. Two overlapped Mariner 4 red and blue filtered images of Mars, together with a "synthesized green," create the "full color" smaller image in-between. Mariner 4's twenty-two historic images revealed for the first time the long history of "ancient impact craters on Mars"—a development that ultimately killed "the Martian legend of a once-inhabited Mars," vigorously promoted in the early twentieth century by Boston brahmin Percival Lowell.

What made *Mariner 4*'s discovery of a "heavily cratered Mars" doubly ironic was the fact that Tombaugh, in addition to his historic discovery of Pluto, from his later Martian telescopic investigations was among a handful of astronomers in the 1950s to successfully predict—well *before Mariner 4*'s 1965 transformative Martian fly-by that July night—the existence of "impact craters on Mars." Ironically (for an astronomer working from the very Observatory specifically created by Percival Lowell to *prove* the existence of Martian life), these craters were skillfully used by "others," in the wake of *Mariner 4*, to begin the political dismantling of the entire "Lowellian Legend" of former (or current) intelligent life on Mars.

But in perhaps the ultimate "Mars/Pluto coincidence," we now find out that NASA's *New Horizons* spacecraft is being deliberately targeted to fly closest to Pluto[32]—six thousand miles "close"—the night of *July 15, 2015*, at 11:49 GM, a mere eleven hours later than the *exact moment*, fifty years before, when *Mariner 4* first flew six thousand miles from Mars, beginning the late twentieth century "excommunication" of a potentially "inhabitable Red Planet."

Eleven hours . . . out of *half a century*. An "error"—across fifty years—of 0.0000295: yup, sheer coincidence.

So, by drawing a *deliberate and precise* (remember, to 0.00295 percent . . .) "half-century link" between the "first-time" exploration of these two "classical" planets, between "the *first* Mars NASA fly-by" in 1965 and this (much politically-delayed) "first Pluto fly-by" in 2015, was NASA trying to "signal" (and, to whom?) its own "insider knowledge" of some kind of "underlying Mars-Pluto *connection*"? To be "revealed" *during* the *New Horizons* fly-by!? A "Mars/Pluto heads-up" deliberately "encoded" in the (obvious) deliberate *timing* of the upcoming NASA Pluto encounter to *precisely* fifty years (plus those eleven "tetrahedral" hours) after the historic legacy of *Mariner 4?*

Precisely fifty years . . .

What could "fifty," as part of a possible "hidden NASA message" in the *New Horizons* fly-by *timing*, mean in the larger context of a possible "*Mars-Pluto Connection*" . . . that would demand a *code* to communicate?

Wasn't it obvious?—

Another NASA ritual.

Given NASA's documented propensity for doing exactly that, over and over again (detailed in our previous, extensive examinations of NASA's stealthy "coding" of key celestial and numerical rituals into its planetary and human missions),[33] I now wondered if the number "50" *itself* might represent not a "period of time" (an orbit), but some kind of NASA *"mythological / symbolic* link" . . . potentially, to some kind of "creation event" or "sacred location" or ancient "god" to which NASA was now, apparently, calling specific, *symbolic* attention . . . in connection with its impending first robotic exploration of "highly anomalous Pluto."

In fact, there was. . . .

In ancient "Mesopotamian" cosmology, Enlil (the Sumerian "chief creator god" of the world's earliest known "high civilization") was also known, in the sacred texts, by his "celestial *number*"—50.

During the subsequent, several-millennia transition of the Sumerian into the Babylonian civilization, the new reigning deity, linguistically, evolved into "Marduk"... still, however, cosmologically identified by the *same* "sacred celestial 50." And "Marduk," by the time of Babylon ... as "chief god, 50" ... also had become identified with—*Mars*.

So, one compelling, *documentable* "Mars-Pluto connection" for *New Horizons* (if treated like NASA has clandestinely treated all its other planetary and lunar missions—*mythologically*), equaled:

An *Enlil / Marduk / Mars / Pluto / New Horizons* Connection–
All by way of "50."
All of which meant . . . "what?"

In Sumerian cosmology / theology, among his other duties, Enlil presided over the *"creation* and guidance of *humankind"* [the same god who later sent the flood (in the "Eridu Genesis")[34] . . . to get rid of his "mistake"].

Could the deliberate attachment of the number "50" to this first "NASA Pluto mission" (by simply, quietly and cleverly, *delaying* it . . . for many years . . . until "the time was right") be a coded signal ultimately relating to "ancient knowledge of humankind's real solar system *origins"*—the story *behind* the ancient Sumerian "Enlil creation myth" of "50"—somehow, involving Mars . . . and *Pluto?!*

Remember Disney, subliminally attaching a "Sirius meme" to *Pluto* in the public mind—with "his new cartoon 'pup' . . . now, forever connected to the newly-found ninth planet?"

Well, in Dogon / Egyptian mythology, "50" was the specific *number* for the *"dog star,"* Sirius, because (the Dogon claimed) of the "fifty-year *elliptical* orbit of Sirius ... and its abnormally "heavy" companion star, *Po Tolo* ... around each other (a Dogon concept held *millennia* before the early twentieth century astronomical confirmation of an Earth-sized, super-dense white dwarf star, orbiting Sirius every *fifty years* . . . "Sirius-B").

Was "50," then, really a multi-leveled, mythologically coded ancient *celestial* reference (again, *secretly* by NASA) — to "a distant planetary epoch of 'the creator *gods*' themselves . . . when, the entire solar system was 'magically' *transformed*" . . . by way of the intervention of the "gods" from a *Type II* civilization?

A process that—via this "sacred celestial number," 50 – would also now seem *directly associated* (because of that "half-century orbital period") with the near-by main sequence A-type star, "Sirius?"

Were we on Earth, living amid the "shattered remains" of a once vast and ancient "Sirius Solar System Renovation Project"—designed to ultimately arrange for a "suitable" planetary environment . . . in this carefully "redesigned," near-by star system — a *designer* planetary environment suitable for . . . *humans?*

Perchance, a carefully-prepared ancient "cosmic '*garden?*'"

Does, in fact, NASA's arcane *New Horizons* "Pluto heads-up"—clearly coded in the extraordinarily precise timing of *New Horizons's* arrival at Pluto in July 2015—quietly foreshadow (to "those who *know*") the imminent, long-awaited NASA publication of *scientific* evidence (from the *New Horizons* mission) affirming that Pluto, indeed, was once part of "a solar-system-wide, ancient ET civilization—quite likely now connected to *Sirius?*"

CREDIT: NASA/THE NEW HORIZONS MISSION

Figure 9. Artificial "geometry" on Pluto? A NASA artist visually "speculates" what the New Horizons Mission might discover.

The most frequent examples of quietly "shouting down the millennia . . . to later epochs," of sending an Important Message down through time to "a select few" (while having the message *ignored* by everybody else), historically, has been to simply "talk in code."

The highly-entertaining Walt Disney Pictures *National Treasure I and II*—through a fictional story-arc beginning with "cracking a series of historical clues deliberately left by the Founders," to "a vast and ancient treasure"—acquainted a lot of folks with an "untaught" historical reality surrounding the founding of the United States . . . a reality that's hardly ever hinted at in school. Before, during, and after the Revolution (just as *National Treasure* indicates), our Founding Fathers did a lot of "talking in code"—encrypting the secrets of the creation of an entire nation in an ancient *symbolic* language (just look at a dollar bill), the deepest translations of which are truly known only to "those in the know" (those belonging to certain "ancient fraternal orders").

A symbolic language, as we've shown elsewhere,[35] that we now know can trace its *ultimate* origins—to Mars.

As *Lost Horizon* was both a remarkable work of entertainment and an important "message film" for the 1930s (just before WWII), so I have come to suspect that Disney's *National Treasure* should be viewed as a uniquely-coded "message movie" for our time, under the guise of "entertainment" and "coincidentally" involving the themes *most* central to our own three decades of Enterprise multi-disciplinary research:

Ancient terrestrial/extraterrestrial civilizations

Ancient advanced knowledge, preserved by those past civilizations

Ancient advanced *technologies* (machines), based on a *real* Physics, archived redundantly from those same ancient civilizations both here on Earth and across the ancient solar system

And, the crucial "terrestrial connection" to all of this: Our proposed ancient ET *ancestor contact,* with the most ancients roots of human civilizations . . . still confined to Earth.

At the end of the second *National Treasure* film, Nicolas Cage (playing Ben Gates, the code-breaking "Indiana Jones"–type at the heart of the franchise) has a fascinating exchange with a (deliberately depicted) "Kennedy-esque" president (Bruce Greenwood) over an "item" in a fictional "President's Book":

> President: Ben, I am curious about that favor I asked you … any report
> regarding what's on page 47?
> Gates: I believe I can help with that, sir.
> President: So, it's good …
> Gates: *Life-altering*, sir.

The fact that this exchange is built around a mythical "presidential entry on page 47," of a mythical "Presidents' Book" of centuries-held "national secrets," definitely should *not* be taken, in this context, just at face value … as merely "entertaining fiction" (the "meta-lesson" of the films themselves). Quite possibly this is a real "code!"

The redundant references to "47" seem a thinly-veiled reference to "1947"—not only the *year* of the most famous "UFO / secret ET technology" case in history (Roswell, repeatedly referenced in the film); but also, via the same "19.47" code, a compelling intimation of the "life-altering" properties of Hyperdimensional / Torsion Field Physics—unleashed in stars and planets at "19.47 degrees" … in a *rotating* planetary / stellar context.

"What's on page 47" could be the ultimate "National Treasure": immeasurably enriching not only the lives of every American, but every human being on Earth … a tantalizing clue to the "new treasure" that might be the centerpiece of the final sequel—*National Treasure 3*, scheduled for release (at this writing) the summer of 2015 … just as NASA could be boggling our eyes and minds with what's *really* on Pluto.

In this *National Treasure code* model, the true nature of Pluto (as a potential "Ancient Archive / HD Generator") would have been deliberately "imprinted in code" into the "Plutonian System" itself … by the Type II Civilization responsible for its "*redesign*." Such that, when a subsequent technological civilization (us?) finally arose again on Earth (even long after "the War"), able to once again translate such a set of stark "Plutonian orbital distances … periods … and other 'Keplerian' numbers" back into the right numerical *code*—the receiving civilization would also recognize (from its own, independent *rediscoveries*) that the ultimate "message" in Pluto's orbit was written in a *Hyperdimensional / Torsion-Field*" code: the language of "the Physics" that *had* to have been used by our proposed "Type II Civilization" to *remodel* the original solar system in the first place!

Pluto's unique, dramatically inclined orbital parameters—compared to the (essentially) "co-planar" solar system inward of Neptune—was, of course, designed (in this "artificial model") to make Pluto immediately stand out as obviously "different" from all the other the inner planets, with its 17.1-degree inclination, 39 AU average solar distance, and 248-year orbital period: unmistakable (to the "right set of eyes") as coded directions to a "higher-order translation."

If Pluto was, indeed, the potential lynchpin in some "Ancient Grand Design" for ultimately preserving and passing along knowledge of "the true history of the remodeled ancient solar system," and over *geological* time, then its creators—our much-discussed Type II Civilization—would have had every incentive, and certainly the means, to make Pluto's crucial role blatantly obvious to any "curious civilizations" that came after.

Including even "cosmically-challenged" descendants, like ourselves.

So, if such an ancient "Pluto code" is real, what should we be looking for via *New Horizons*—as physical *evidence* supporting these radical ideas . . . potentially detectable (if we know "what" we are looking for) to the wide variety of instruments carried aboard the *New Horizons* mission?

Proceeding on the logical assumption that our proposed Type II Civilization redundantly encoded "HD information" in *many* aspects of its "redesigned" ancient solar system (as, preserving "the Physics" was clearly one of its highest demonstrable priorities), the simplest place to start with regard to Pluto, as noted earlier, would be to look—à la *National Treasure*—for a "code," preserved in the most obvious location possible—Pluto's basic *orbital parameters*.

Let's begin, then, with that fundamental "248-Earth year orbital period" for one "Plutonian year"—easily established by comparing Tombaugh's initial series of discovery photographs to Pluto's incremental movement on photos taken later. That number—248—is also, astonishingly (to *three significant figures!*) exactly the same number as Pluto's orbital eccentricity[36] (elliptical departure from a "perfect" circle, as it orbits the Sun) along that same "248-year path":

$$P = 248 \ldots Ecc = 0.248$$

The odds against these two numbers being *the same* (and, to three significant figures) "by sheer coincidence," were (literally) astronomically *against*, because the two parameters—orbital period and orbital eccentricity—were (as mathematicians would say) "totally decoupled"; neither one *physically* depending on the other for *any* specific value, from zero to "infinity."

Yet here, with Pluto, those two *totally independent* physical orbital parameters were, somehow, numerically *identical*. How come!?

Lacking any known natural "drivers" for such an extraordinarily low-probability event, the only plausible alternative is that the Pluto year/eccentricity "coincidence" was *designed,* part of the deliberate Pluto "heads-up."

So, what was *intrinsically* "important" about 248? Why not an *exact* "250" ... or some other randomly chosen period?

Well ... "248" happens to be the "lowest order" product of the multiplication, by "two," of each *succeeding* number in the same string ... *starting* with the number "2":

"Two times *two* equals four ... two times *four* equals eight, etc."—

2 4 8.

To a first order, the fundamental "intent" behind selecting "248" would seem to have been to communicate both a *geometric* progression (rapid "amplification" of a starting number—2 when multiplied by 2 ...), simultaneously, linked with a *harmonic* sequence (indicating "frequency") per the same *doubling* of the preceding number–

2 ... 4 ... 8.

This strongly implies that "the most important 'something' to note about Pluto"—when we could physically reach out to it—might be confirmation of some kind of "*harmonic* ... *amplifying* process" occurring there. A major clue to what that "process" might be would seem to lie in the additional orbital "coding" (in our model), inherent in the third listed orbital parameter: Pluto's average distance from the Sun... a little more than 39 AU.

And 39, of course, is *twice* "19.5"—strongly implying (in the "code") that the "harmonics" we should be looking for will *not* be "electromagnetic," but will lie in Pluto's invisible (to "normal" detectors) changing "*HD/torsion field*" ... which creates, in rotating planetary and stellar objects, a distinct "energy upwelling signature"—at "19.5 degrees."

Perchance, an artificial "*torsion-field generator*"—carefully placed somewhere on Pluto (at 19.5?) or elsewhere in the Plutonian system (on one of Pluto's moons?) by our proposed "Type II Civilization" as a deliberate "beacon" to further mark "its crucial solar system Archive!"

The first mainstream scientist (that I know of) to seriously propose that Pluto might, in fact, be "some kind of super-advanced, '*ET mega-mechanism*'"— attempting to scientifically explain, within our 1960s aerospace knowledge-base, the mounting "anomalies" surrounding Pluto's increasingly "physically impossible" *confirmation*—was a gifted physicist working for the Hughes Aircraft Company Corporate Research Laboratory, Dr. Robert Forward.[37]

Bob's fascinating "grand Pluto idea" (published long before our curious encounter, in the December 1962 issue of *Galaxy* magazine, as "Pluto: Door-way to the Stars")[38] was that Pluto might *not* be "just another member of the outer solar system." Forward's suspicion was based on *decades* of continuing failure (even into the 1940s and '50s) of the largest telescopes on Earth (like Mt. Palomar's famed "two-hundred-inch Hale Reflector") to detect any hint of a *disc* at Pluto—corresponding to "a planet matching Lowell's pre-discovery gravitational estimate of 'six or seven Earth masses.'" Forward's grudging conclusion: "Pluto," in fact, might not be a "planet" after all (shades of the forty-four-years later IAU decision)!

Bob Forward's counter-proposal (buttressed by a lot of "relativistic phys-ics") envisioned a radical alternative—that, instead of being another natural planet, Pluto might, in fact, be:

"A deliberately-created, high-tech ET artifact!"
Left as a Gift . . . from a hypothetical "Galactic Federation."
Perchance . . . a "relativistic Galactic transport mechanism"—designed
    as a deliberate "intervention" by ET
To "help us toward the stars" . . . when we were "ready" (practically
    defined as "being able to physically reach Pluto").

(And some folks think *I* have an "overactive" scientific imagination.)

Forward calculated that such a far-reaching "ET interstellar transport technology" could be built around "a relatively 'small quantity' of degenerate (white dwarf-density) stellar material" crammed into a donut-shaped con-figuration, physically "about the size of Mercury"—but containing around *six times Earth's mass.*

By swirling this degenerate matter around the "donut" at almost the speed of light (!) until it (à la "Relativity") visibly "bent space," the resulting "arti-ficial gravity" of such a whirling Ring—acting uniformly on every atom of any vehicle using it, Forward calculated—was capable of gravitationally accelerating a manned spaceship (because of its relativistically-enhanced mass) to close to that same speed (the speed of light) *in a matter of seconds,* after which, the ship (and its occupants) would coast between the stars for years before arriving at its destination. This would allow—all based on "known physics"—genuine human interstellar travel (when we physically *reached* Pluto) *if* (and, it was a big "if"):

The same "super-advanced ET civilization" had thoughtfully placed an *identical* "relativistic ring" at the other end of each potential "interstellar journey" from the solar system—wating in orbit around all the nearest stars—which, when properly approached geometrically, would gently slow the accelerated spacecraft back down to "interplanetary speeds" for a successful planetary rendezvous at the other end of such "relativistic interstellar journeys": a "Galactic Transport *Network*" (ah . . . *Stargate: SG-1* anyone . . . again?).

In part, Forward's exotic "Plutonian gravity generator / galactic transport node" was concocted, in 1962, as a somewhat "desperate" conventional physics attempt at an explanation of Pluto's (apparent, under some interpretations of Lowell's bafflingly accurate prediction) "*variable* mass" (seriously!); Forward was as bothered by Lowell's (apparent) "luck" in successfully predicting "just by accident" the pre-discovery location for Pluto, despite his serious mathematical errors, as a lot of other folks were:[39]

> This [the Pluto confirmation, initially] looked like another fantastic piece of mathematical theory by Lowell. However Brown reviewed the data which Lowell had used and showed that *there was no way that he* [Lowell] *could have made the correct* [Pluto positional] *prediction based on the* [anomalous Uranus and Neptune] *data.* [emphases added]
>
> —J. J. O'Connor and E. F. Robertson,
> "Mathematical Discovery of Planets"

Thus, Forward's amazing "ET Pluto Gravity Ring" had the potential—within the realm of "known" 1960s science—to (finally) physically answer the serious "Pluto discovery discrepancies."

A prime feature of such a Ring would be that, when oriented at certain angles relative to the other planets in the solar system, it could, theoretically, *vary* its innate long-distance gravitational effect (which, unlike true gravity, was not "spherically symmetrical") on these other solar system objects—in a process akin to a classical "science fiction tractor or presser beam" effect.

But, at other times—when this *non-spherical*, "relativistically-created gravity machine" was *not* aimed at those planets—nothing.

Forward thus theorized that this "*varying* gravitational attraction"—directly as a result of General Relativity and his "artificial Pluto's" inexorable precession—could be the reason why, when Lowell's ephemeris discrepancies for Uranus and Neptune were compared with later efforts, the two sets of observations *didn't match*; Pluto's current miniscule physical gravitational attraction on Uranus and Neptune, because of the continuing precession of the Ring in the intervening decades, and the associated alteration of its "gravitational attraction geometry," had simply (in Forward's theory) inevitably *decreased* any observable effects on Uranus and Neptune.

Bob Forward died in 2002, about a year before the first, startling "inertia-*is*-capable-of-being-changed" empirical results from our own ten-year series of Enterprise "HD/Torsion Field" experiments began coming in; these experiments started with the acquisition of remarkable Enterprise Accutron data from the rare Venus Transit of the Sun[40] that we observed from Coral Castle, Homestead, Florida in 2003.

This was followed, over the succeeding decade, by on-site Enterprise direct measurements of the changing terrestrial background torsion field from a variety of other global locations—including a list of "ancient, sacred sites"—culminating with the dramatic and highly complex inertial changes measured directly in the path of an eclipse of the Sun, as seen from Albuquerque, New Mexico in 2012.

From this decade of archived Enterprise HD/torsion background data, it has become clear that local "aligned and spinning celestial objects" (like the Sun, Moon, and transiting/eclipsing *planets* like Venus) all create *measurable* changes in the inertia (mass) of appropriately designed torsion-field detectors—when those detectors are placed directly within *geometric* planetary/moon alignments/transits/eclipses.

Further, these *observed* inertial changes are more than *one hundred million times* stronger than distant effects predicted by Forward's proposed "inertial/gravitational ring," based again on Einstein's General Theory of Relativity.[41] This makes them *easily* detectable by anyone using the Enterprise Mission's unique Accutron torsion field detection system, seen here against a portion of the astonishing HD/Torsion solar eclipse data we acquired with this system in 2012.

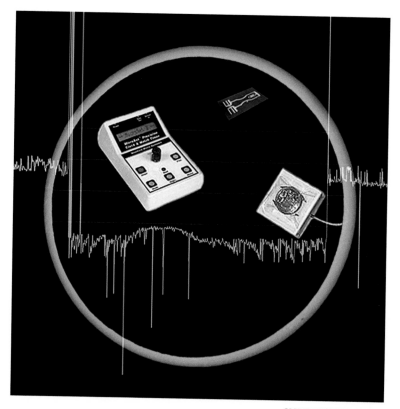

Figure 10. Schematic of The Enterprise Mission's "Hyperdimensional/Torsion Field" Accutron measurement system—depicted against an image of the May 5, 2012 annular solar eclipse, and the actual Accutron "HD/torsion data" acquired for the first time by Enterprise during that eclipse.

Torsion conditions that will certainly also exist, based on these extensive Enterprise celestial torsion observations, for *New Horizons* at Pluto. Created by the *synchronous* Pluto/Charon 6.4–day rotation/revolution, and further synchronized with the constantly moving/constantly *resonating*/constantly *eclipsing* four (currently known) additional Plutonian moons, such a specifically-designed, "resonant torsion *system*" (in the model) could not help but produce unique, time-varying *torsion field-effects* all across the solar system—including on distant Earth. These would peak (amplify its generated torsion field pattern *geometrically*) every 124 years, as Pluto's -120-degree obliquity to its own orbit caused the "generated, aligned, *opposing* twin polar torsion-beams" to intersect the Earth twice every Plutonian Year, in a *recurring*, 248-years-long Plutonian total "Hyperdimensional 'astrological'" pattern.

This model is increasingly supported with every new official Pluto observation, the latest example being Hubble's detection[42] (beginning in 2006) of those four *additional* co-planar Plutonian moons—*Styx, Nix, Kerberos*, and *Hydra* (discovered decades after *Charon,* Pluto's previously single known moon, fully half the size of Pluto, which was initially recognized on photographs taken by the US Naval Observatory in 1978).

The four additional Pluto "Hubble moonlets," though insignificant compared to *Charon* (merely a few tens of miles across, against *Charon's* 740 miles), turned out to possess extraordinarily "harmonic" *3, 4, 5, 6 orbits* relative to both Pluto's own rotation and *Charon's* identical rotation/revolution.

In the Enterprise model, these precisely "tuned Plutonian satellite parameters" were, once again, part of a deliberate ancient design to reinforce its possible primary, artificial "harmonic torsion *message*"—explicit in the Plutonian system's unique "harmonically coded" *orbital* motions.

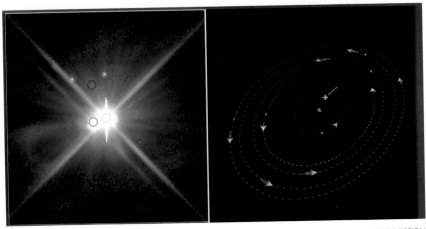

CREDIT: NASA/THE NEW HORIZONS MISSION

Figure 11. Hubble Telescope Pluto image, compared to a schematic of the currently known Pluto "system."

Here (below), by way of example of the radically different types of "torsion readings" that might be (secretly) returned from *New Horizons* to NASA as the spacecraft flies through the Plutonian system, are a few hours of recent Enterprise Accutron "occultation torsion data" of the Sun, a record of the dramatic background inertial field changes (in our hypothesis) that occurred in the Accutron detector during the "ultimate" planetary alignment in May 2012.

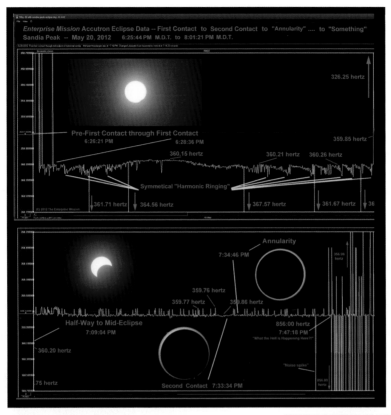

Figure 12. Enterprise Mission "time-versus-frequency" record of the dramatically changing "HD/torsion inertial field," measured by monitoring the Accutron tuning fork frequency changes occurring during the May 20, 2012 annular solar eclipse.

As can be seen, even before "First Contact" (above, far left), the Enterprise Accutron detector registers the lunar occultation of the solar *chromosphere* by responding with an abrupt frequency change of its internal tuning fork, before a dramatically *different* torsion signature then appears just seconds later— marking the formal beginnings of the eclipse itself by the occulting Moon.

As the eclipse continues to progress (see image, left to right), and the Accutron tuning fork frequency continues to dramatically change, it eventually became clear (in the course of three years of "post-Eclipse analysis" we have performed on this data, so far) that what we are seeing in the Accutron graph was nothing less than a "2-D torsion field map" *of the interior of the Sun's and Moon's* various "torsion-generating/occulting layers"! These changing frequency readings were created, we hypothesized, as the Sun's "solar plasma torsion-generators" were being successively covered and uncovered by the constantly moving, rotating interior layers of the eclipsing Moon, resulting in a set of changing, interfering torsion frequencies during the Sun and Moon's constantly *changing* geometric alignment!

In our provisional model, these differentially-rotating plasma layers inside the Sun—of radically-differing temperatures, ionization states and densities, from the surface to the core—were each, apparently, generating individual dominant torsion-field vibrations in the "aether" unique to that particular rotating plasma layer; the eclipsing Moon—with its own, equally-layered interior geology—was apparently, via its solid-body rotation, creating a series of *interference torsion frequencies,* as both rotating celestial objects' individual torsion-field patterns "beat" against each other.... Serendipitously, by sensing this *inertia-altering* process with our Accutron's miniature tuning fork (which, through its rapidly changing inertial response to the changing torsion field, had changed its fundamental frequency of vibration to match), we had produced the world's first (public) "2-D torsion cat-scan" of the *interiors* of both the Sun and Moon (below)!

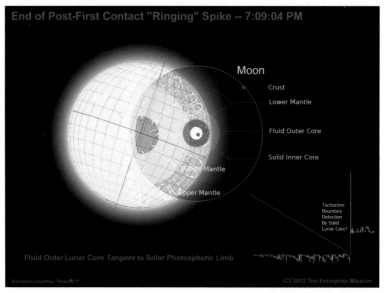

CREDIT: THE ENTERPRISE MISSION

Figure 13. Abrupt Accutron tuning fork frequency change (bottom right) correlates, in this cut-away diagram of the interiors of the Sun and Moon, with the moment the Moon's inner, solid core (small white circle inside the larger orange one) tangentially occults the Sun's deep, swift-moving "tachocline layer" at the base of the outer convection zone – providing totally independent torsion confirmation of the remarkable accuracy of current interior models of the Sun and Moon.

The same capability now existed within NASA (and its *New Horizons* mission) to not only detect any "torsion beacons"' that our great, great, great, great ancestors may have left to mark the proposed Archive's location on Pluto—but the simultaneous ability to create 2-D "torsion maps" of the *interior* geological structures of both Pluto and Charon.

All that would be needed would the "right" *New Horizons* trajectory through the Plutonian system, and the right " torsion detection instrumentation" on board.

From the looks of it (see image), *New Horizons* is, in fact, going to be shot (at thirty thousand miles per hour) through the perfect "eye of the needle" during its several-hour Plutonian fly-by, navigated to the "ideal position" to conduct direct, *secret* "HD/Torsion Field" measurements of the entire Plutonian system—along with its (much-advertised) "conventional" planetary science.

This will include torsion readings acquired from *inside* two very carefully-arranged "satellite eclipses" of the Sun, by Pluto and Charon, as *New Horizons* is directed to fly diagonally across their extremely extended shadows . . . within hours of the spacecraft's "closest-approaches" to the two eclipsing objects themselves. Potential torsion measurements that, from our own previous Enterprise experiments, we now have little doubt will quietly confirm profound "hyperdimensional/torsion-arranged signatures" emanating from the entire Plutonian system . . . directly (if clandestinely) detectable by *New Horizons*.[43]

Figure 14. NASA artistic vision of a series of "cryo-volcanoes" on Pluto, with the ultra-thin atmosphere visible in "forward-scattered light" from the ~39 AU (twice 19.5 . . .) distant sun.

Given that astronomical "HD/torsion effects" (in our physics model) are the direct result of intrinsic *rotation,* coupled with (orbital) *revolution* of two or more massive celestial objects around each others, all modulated by their constantly changing relative geometric orbital *alignments* . . . the fact that *New Horizons* will pass almost *vertically* through the "common orbit plane" of *all* of Pluto's "suspiciously" co-planar and "harmonic" moons (see image), will provide a unique HD geometry for measuring the predicted "interference torsion field effects" of Pluto's current polar axis (and concomitant equatorial satellite orbital) orientations—including, as previously noted, carefully arranged HD/torsion measurements to be conducted in passing directly *through* Pluto and Charon's respective "HD/torsion *shadows,*" which, from the spacecraft's viewpoint represents two back-to-back eclipses of the distant sun).

CREDIT: NASA

Figure 15. An artist impression of an eclipse of Pluto by its single large moon, Charon. As empirically determined by the Enterprise Mission, terrestrial eclipses have measureable "HD/torsion field effects." The geometry of the New Horizons fly-by through the Plutonian system has apparently been carefully arranged to allow four "artificial eclipses" during the Encounter—into the shadows of both Pluto and Charon, as seen from the sun and Earth . . . allowing NASA unique "in situ" HD/torsion measurements of the Plutonian system.

Again, from our own repeated Enterprise Mission Accutron HD/Torsion Field measurements—especially of the Annular Solar Eclipse that passed directly over Albuquerque on May 20, 2012—it is obviously, from our perspective, no coincidence that NASA has carefully planned for the *New Horizons* spacecraft (in those familiar words from *Forbidden Planet*)—"to arrange for its own eclipses."

Only in this manner can *New Horizons* "torsion scan" the *interiors* of Pluto and Charon, to discern if there are, indeed, any anomalous *geometric* features underground, potentially associated with an "artificially-designed ET Archive," *hyperdimensionally* marking" the precise location of those surviving ancient Libraries. Simultaneously, it can test the possibility that Pluto—in addition to its potential role as a "physical time-capsule/Archive" from the ancient solar system—may also have been carefully *placed* in its unique solar system orbit to act as a deliberate, high-angle "planetary HD/Torsion Field beacon/generator"—as part of the carefully interlocking, "Hyperdimensional/Torsion redesign" of the entire solar system.

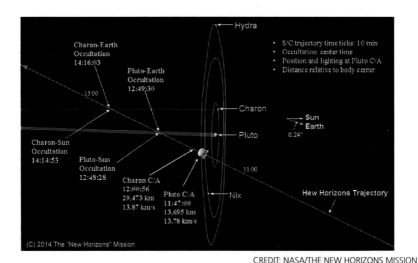

CREDIT: NASA/THE NEW HORIZONS MISSION

Figure 16. Top-down view of New Horizons ~90-degree fly-by geometry through the Plutonian system; note the four scheduled eclipses of the sun and Earth ("occultations") this carefully chosen trajectory allows by Pluto and Charon—insuring unique HD/torsion insitu measurements of both large orbiting bodies . . . if NASA deigns to tell us.

Was Stern's *New Horizons* mission (another name replete with "ancient Egyptian *'reincarnation'* overtones"—and we *know* how NASA *loves* "Egyptian overtones") really a mission to ultimately, symbolically *reincarnate* our own cultural planetary awareness of our *real*, "ancient, Type II solar system heritage"?

Remarkably, Pluto's own "coded" inclination (in the model) of 17 degrees to the rest of the inner planets' orbits, perfectly matched the ancient Egyptian invocation of "17" as a number symbolic of "renewal and resurrection."[44] Was *New Horizons* to be the instrument of that "resurrection" by simply, at the "right time" (judged, again, "ritually"), telling us the *truth?* And, were "mid-2015" and the upcoming amazing *New Horizons* Pluto fly-by the beginnings of "That Time?"

OK, it's a great theory (I hear you saying), but where's the evidence?

Again, "in plain sight"—in NASA's (and other space agency) official archives: if you know "how" to look.

Out of all the NASA (and other space agency) robotic expeditions sent across the solar system over the last half-century, most of the "small objects" those missions have successfully encountered—always carefully labeled as just "another comet or asteroid mission" in all news coverage—actually appear, from quietly "leaked" information coming out of NASA, to be nothing less than visits to long-abandoned, ancient *ET spaceships!*

The most blatant, recent example of this "in plain sight" gambit is the European Union's (ESA) unmanned *Rosetta* spacecraft mission, placed into "loose orbit" of comet 67P/Churyumov-Gerasimenko August 6, 2014—a little less than a year before the Pluto *New Horizons* fly-by.

Eventually *Rosetta* would attempt to land a smaller unmanned probe, named "Philae," on the comet's surface. At this writing, high-resolution *Rosetta* images are coming back from less than thirty miles off comet 67P, as the spacecraft slowly cruises through a bizarre set of (ritual?) equilateral *"triangular* approach orbits" (see image).

Mission: to survey a comet

The European Space Agency's Rosetta spacecraft will carry out tough manoeuvres once it arrives at comet 67P/Churyumov-Gerasimenko

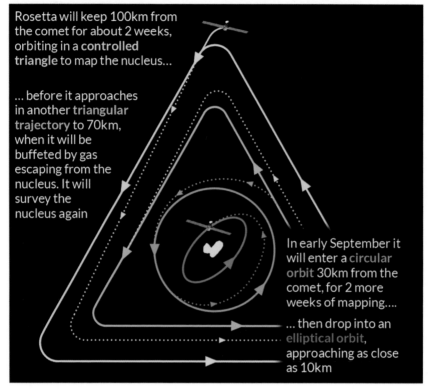

Rosetta will keep 100km from the comet for about 2 weeks, orbiting in a **controlled triangle** to map the nucleus...

... before it approaches in another **triangular trajectory** to 70km, when it will be buffeted by gas escaping from the nucleus. It will survey the nucleus again

In early September it will enter a circular orbit 30km from the comet, for 2 more weeks of mapping....

... then drop into an elliptical orbit, approaching as close as 10km

CREDIT: ESA

Figure 17. Set of unique triangular ("tetrahedral" . . . "ritual?") final approach orbits, carried out by the unmanned *Rosetta* spacecraft as it moved closer to Comet 67P/Churyumov-Gerasimenko in August/September 2014.

In the 67P "approach close-ups," at certain angles, the object was seen by some as taking on a cartoon-like "rubber ducky" resemblance (see image).[45]

Figure 18. *Rosetta* images Comet 67P/Churyumov-Gerasimenko "from the head, looking down toward the larger 'body'" with its high-resolution Osiris camera, from about 100 miles away in early August, 2014 . . . prior to establishing an orbit of 67P. Note the remarkable "symmetries" and "right-angle features" visible all across the two-lobed surface in this image. Such "geometric (even if badly eroded . . .) symmetry" strongly hints at an artificial origin for these parallel and right-angle geometric features, if not this entire "comet."

Remarkably *symmetric*.

Another perspective image, acquired a few days later during *Rosetta's* continuing approach (see image), clearly revealed a series of "suspiciously *geometric*" protuberances all along the limb of both the comet's "lobes."

Figure 19. Different orientation of Comet 67P/Churyumov-Gerasimenko, relative to the *Rosetta* spacecraft. The "head" in this Osiris image is to the right; the comet's "body" is to the left. More "rectilinear" and "multileveled" heavily-eroded geometry is also readily apparent across the surface, "ordered" geometry impossible to produce "naturally" . . . .

A close-up of the larger "lobe" in this *Rosetta* spacecraft *Osiris* camera image of "comet 67P" (see image) reveals the remains of what appear to be eroded, *gigantic buildings* on its apex; a series of right angle, quasi-parallel, highly eroded "decks" can also be seen below this "summit"—indicating a porous, large-scale "honeycombed" interior *under* the remaining surface layering.

Figure 20. Closer view of the apex of the larger "lobe" of Comet 67P/Churyumov-Gerasimenko, revealing geometric surface structures resembling "ancient ruins and the exposed edges of eroding decks. . . ."

A third "zoom" (below) leaves little doubt. . . .

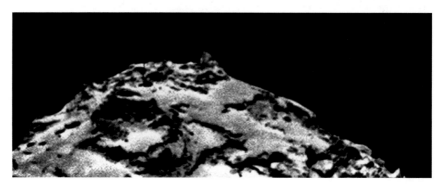

Figure 21. The closest view, from this angle, of the collection of "eroded ruins" arranged around the apex of the "comet's" larger lobe; the presence of such symmetry and geometric "order"—on a "rapidly evaporating, chaotic icy surface . . ."—allows only one reasonable conclusion: this geometry was artificially created—and not from "ice" . . . and must be, because of the severe erosion evident, also extremely old . . . . .

In another, even closer Osiris oblique camera view (of the smaller lobe or "head"—see image), a veritable "ancient archaeological *complex,*" equally heavily-eroded, is stunningly visible along its apex too.

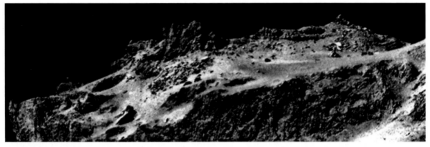

Figure 22. In this later *Rosetta* "NavCam" view, what appears to be the remains of an ancient, heavily-eroded "archaeological complex"—with small clusters of much more "modern" structures in between—stands atop another apex of the "comet's" smaller lobe.

And, an even closer view (below) reveals an exquisitely geometric, multi-leveled, clearly artificial wall, stretching across another apex of the smaller lobe.

Figure 23. Another, even closer NavCam perspective on a different "edge' of 67P's smaller lobe, reveals a massive "wall"—composed of highly complex, geometrically-arranged "substructures"—obviously built by "someone."

Only exceeded by this new *Rosetta* NavCam view below—when the spacecraft (on September 10, 2014 drifted within eighteen miles of the comet.

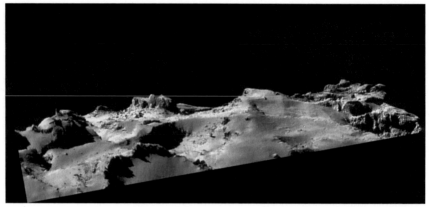

Figure 24. Along another edge of Comet 67P/Churyumov-Gerasimenko's smaller lobe, in this subsequent NavCam view, an array of massive, obviously eroded individual buildings appears . . . each on the scale of several standardized "terrestrial city blocks"—in three dimensions. . . .

And photographed "the remains of visibly ancient, massive, highly-eroded . . . *individual,* clearly artificial structures" on Comet 67P / Churyumov-Gerasimenko (below).

Figure 25. A close-up of two adjacent structures seen in Figure 23, reveals their radically different architectural designs—indicating either "heavily-shielded fortifications" . . . or, massive, sealed "arcologies" for housing tens of thousands. . . .

Equally significant (besides their sheer existence) was the fact that these artificial "comet" structures appeared eerily *familiar*—as seen in this scaled comparison of Comet 67P/Churyumov-Gerasimenkowith downtown Los Angeles.

Figure 26. Artist's scaled comparison between the City of Los Angeles and Comet 67P/ Churyumov-Gerasimenko. Note that several downtown skyscrapers are required (in terms of volume) to equal one "mega-structure" on 67P.

This confirmed the remarkably "human" geometry and scale of the potential *architecture* being blatantly imaged on the surface of "67P," further reinforcing the impression that this increasingly bizarre object was definitely *not* just "a comet," but quite likely another example of an "ancient, longabandoned interplanetary *derelict*"—either a ship or an entire interplanetary *habitat,* still venting into space (below) wisps of its once massive internal atmosphere, as it periodically orbitally approaches (and warms from increasing exposure to) the Sun, every 6.5 years.

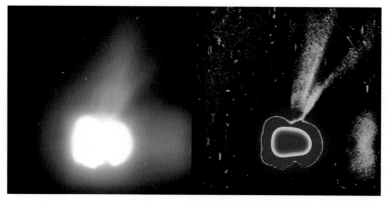

CREDIT: ESA

Figure 27. Escaping water (at a rate of about one glass per second) forms spectacular twin jets rushing into space from the "neck" connecting the two lobes of the 67P/ Churyumov-Gerasimenko "comet" in these approaching July, 2014 Rosetta images. This "collimated jet phenomenon" is a stubbornly baffling feature of all "comets" selected for close-up spacecraft reconnaissance over the last decade; the jets' pervasive existence has created major (if still generally unacknowledged) difficulties for the traditional mainstream comet model, of "surface pockets of cometary ices, flashing into vapor under the increasing heat of the Sun." Despite intensive imaging of multiple objects, a surface source for these widespread gas and dust emissions has not been found on any "comet" visited . . . reinforcing the Enterprise Mission's radically different "ancient spacecraft" model, that inherently predicts an interior, mechanical source (passageways or elevator shafts) for passively collimating such widespread "anomalously linear gas jets."

Alternatively, the Europeans could have "stumbled upon" the first astonishing *direct* evidence of Tom Van Flandern's "exploded planet hypothesis"[46]—a genuine, miles-wide, continental *fragment*, complete with "the ruins of a once massive planetary *city* still anchored to its bedrock," blasted into space when the planet underneath it, literally *exploded* . . . leaving provocative evidence, in the shattered geometric form of Comet 67P/Churyumov-Gerasimenko and its colossal, surface "mega-structures," of the extraordinary power of the Type II Civilization which once remodeled an entire solar system, and the equally potent forces which, much later, radically, violently, attempted to destroy it. Thee *Rosetta* images made the possibility of a deliberate "Ancient Solar System Archive—carefully preserved on high inclination Pluto" (or, somewhere in the rest of the Plutonian "system"), in our opinion, even more deserving of serious consideration.

After seeing the amazing "approach" images to "67P,"[47] the fact that this seminal ESA mission is named "Rosetta" should be another, major clue regarding its own possible "*real* mission"—certainly to anyone who has followed NASA's own deeply "Egyptian" predilections over the last half-century. *Rosetta*—openly named after the "game changing" breakthrough in Egyptian hieroglyph translation of an "extinct terrestrial civilization" in the nineteenth century—can only be viewed as an even more telling designation *now*. This mission had the potential to reveal as much about the Type II "extinct civilization" that left "67P" for us to find . . . if *Philae* successfully touches down in the "right" place . . . as the original Rosetta Stone revealed to the nineteenth century about the Ancient Egyptians.

But, *Rosetta* isn't the only government space mission to have secured astonishing, close-in measurements of an "ancient, outgassing *spacecraft*"; the close fly-by of Comet Hartley 2, in late 2010, by a NASA spacecraft called *Deep Impact* (re-targeted from its previous, successful "impacting mission" to another comet), provided amazing, even earlier confirmatory data for our Enterprise "remodeled Solar System" model.

On the highest-resolution *Deep Impact* NASA images of Hartley 2 (see image), the true nature of this misnamed "comet" was stunningly revealed . . .

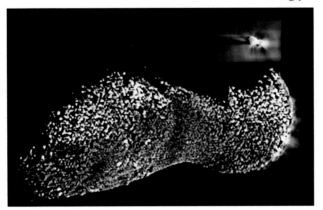

CREDIT: NASA AND THE ENTERPRISE MISSION

Figure 28. The true nature of "Comet Hartley 2" (1.25 miles across) is starkly revealed in this "leaked," high-resolution Deep Impact NASA image, taken during the spacecraft comet fly-by in 2010; miles of highly organized, internal "building-like" geometry are nakedly exposed through the (now missing—from constant micro-meteorite erosion) outer "hull"—just one hint at how truly ancient this derelict must be . . . . Narrow gas jets, vestiges of the remaining internal atmosphere still escaping into space, can be seen along the "ship's" right-hand limb (a wider view of the "collimated geometry" of this continued outgassing can be seen in the smaller, deliberately over-exposed inset image). The presence of similar collimated jets, gushing from the "comet's" dark (and frigid shadow—inset), effectively demolishes the "surface warming theory"; where would the energy to vaporize surface ice lying in that shadow (in "the standard comet model") come from in the dark? The internal "gas release model," with remaining water vapor and other identified constituents venting along deep internal passageways, fits Deep Impact's observations of Hartley 2's "nightside jets" perfectly . . . .

. . . as a genuine, one and a quarter mile-long *ancient interplanetary spaceship!*

Its long-since eroded "hull" disclosing cubic *miles* of former interior, 3-D "rectilinear geometry," instantly revealed Hadley 2's "true, *architectural* identity."

So, does the upcoming NASA Pluto fly-by provide any additional opportunities—besides the (for most folks, too arcane) HD Physics— for testing our sweeping "redesigned ancient solar system" hypothesis?; of detecting, during the month-long, *New Horizons* planetary encounter, other scientific evidence that could *confirm* Pluto's potential "unique role" in that hypothesis—as a *deliberate,* ancient solar system Archive?

Remarkably, we may have a pretty exact "preview" (again, courtesy of NASA's own publicly available interplanetary data archives) of just what *New Horizons* could allow us to *see* at Pluto, that would indicate the possible presence of "intelligence"—in the form of an earlier-imaged, "distinctly *artificial-looking,* planetary surface *geometry*" blatantly visible on "quarter-century-old" NASA vidicon images of Neptune's largest moon, Triton.

When NASA's unmanned *Voyager 2* flew by Neptune in August, 1989,[48] it gave us an eerie foreshadowing—a quarter-century before—of what we might experience, in much higher fidelity and resolution (to be expected, in the course of the intervening quarter-of-a-century of technological progression), during a future Pluto fly-by; this image from *Voyager 2* below is one of the few close-up *color* (composited) vidicon views of Triton, recorded during the 1989 *Voyager* spacecraft encounter.

CREDIT: NASA

Figure 29. In this 1986 Voyager 2 close-up, taken near the south pole of Neptune's single large Moon, Triton, we can see a fascinating diversity of totally unexpected geometric forms—apparently (from separate Voyager spectral measurements) buried to varying degrees beneath a surface of solid, nitrogen ice. This striking geometry is increasingly revealed (presumably because of a thinning of the overlying ice cover . . .) as you proceed toward the right-hand side of the image. This extraordinary right-angle surface geometry on Triton does not look "natural."

On this Triton mosaic (above), at the visible diagonal boundary between the newer nitrogen/methane "snows" on Triton's southern polar cap (left side of image) and the older "dirty brown/pink" nitrogen ice (from solid carbon mixed in) plains surrounding that polar cap (right side of image), one can see an array of obvious, buried and semi-buried "strikingly *rectilinear* geometries" (see close-up below).

CREDIT: NASA AND THE ENTERPRISE MISSION

Figure 30. Close-up of the right-hand section of Figure 28, revealing in higher detail the remarkable, repeating architectural geometry . . . still half-buried in the ice . . . blatantly visible in the polar regions of Triton.

Most planetary scientists, after Voyager's close-up environmental observations of Neptune's largest moon, expect Pluto geologically to be a close-analog to Triton. Will that extend to the presence of "half-buried ruins" as well?

Referred to euphemistically by NASA as "cantaloupe terrain," the extraordinarily *geometric* formations actually look like something striking familiar, but categorically "impossible" by all mainstream NASA "science"—"flooded" *archaeological ruins,* still visible after untold *millions* of years, half-buried in the frigid nitrogen "solid atmospheric surface" of Neptune's largest bizarre "retrograde moon."

At Pluto—primarily (in the Enterprise model) because of Pluto's dramatic 17-degree orbital inclination—if there are similar "ancient, ET ruins" visible to the *New Horizons* cameras on its surface, they may be the *best-preserved of any in the solar system*—a direct effect of the (calculated) drastically lower rate of *background* micrometeorite erosion expected on Pluto, compared to the rest of the inner solar system.

Erosion which should drop *precipitously* every time Pluto orbits above (or below) the solar system's central plane (which, due to Keplerian celestial mechanics, is *most* of the time—see diagram, below).

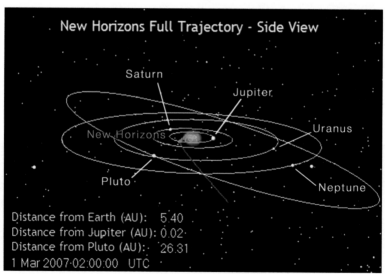

Figure 31. Perspective computer view of the outer solar system, with the New Horizons' trajectory marked in red. The "time hack" is March 1, 2007, just over a year after the spacecraft's launch to Pluto—which can be seen far above the general solar system plane . . . moving downward, to its July 14, 2015 rendezvous with the red line representing NASA's first mission to the controversial "ninth planet."

Thus, any preserved "Plutonian ruins," compared to those found so far in the rest of the solar system, might have been protectively frozen for millions of years . . . *under* "a million, billion tons" of solid nitrogen. Ironically, they would be frozen in *pristine* condition by the side-effects of the same, catastrophic ancient cosmic war that *murdered* (in our Enterprise model) the rest of the inner solar system, allowing a greater likelihood of preservation of "ancient Type II ET science and technology" located *inside* the Plutonian ruins themselves.

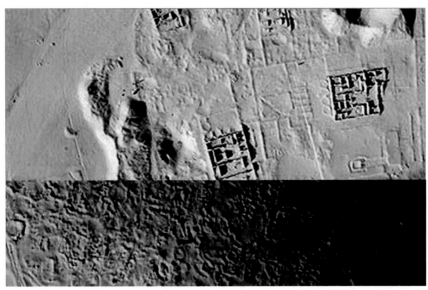

CREDIT: NASA AND THE ENTERPRISE MISSION

Figure 32. Archaeological comparison between the "half-buried ruins at Armarna, Egypt . . . and the morphologically identical "half-buried ruins" on Triton. The only meaningful difference is that one set of ruins is half-buried in sand, and the other is half-buried in nitrogen ice. . . .

A celestial "Shangri-La," preserved against a natural Galactic catastrophe, a local stellar cataclysm, or a catastrophic interplanetary war waged with planet-killing "HD/Torsion Weapons"—a "safely-preserved *archive*" of the real history of "what this Type II civilization had 'rearranged' for our 'creation'"— when we were culturally "prepared enough" (à la "Brookings") to finally know.

Conceivably, under the frigid, miles-deep layers of solid nitrogen ice projected for Pluto's southern hemisphere, there are preserved, just waiting, entire, intact libraries of *first-hand* images and video of the millions-of-years-old true "Galactic history of Civilization," starting from *the very beginnings* of the solar system, if not long, long before.

This hypothesis, judging from the vivid examples at Triton, will soon be verifiable at Pluto via the *New Horizons* mission, and within reach of our subsequent twenty-first-century human and robotic space technology to *physically* explore within a generation.

But only if we *really* want to *know more.*

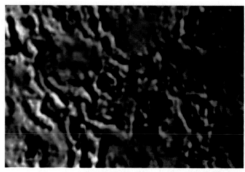

Figure 33. Enlargement from the lower "Triton-half" of Figure 31, reveals striking details of the obvious partially-buried architectural ruins Voyager 2 imaged on Triton . . . a quarter-of-a-century ago. . . . Presently, there seem to be at least three national space agencies—NASA (United States), ESA (Europe), and CNSA (China)—that have all recently, "coincidentally," released official spacecraft imaging data indicating "a former, incredibly advanced intelligent presence in the solar system."

CREDIT: NASA AND THE ENTERPRISE MISSION

NASA's "creeping disclosure" gambit seems to consist of releasing increasingly blatant *Curiosity* images of the smashed and buried "ruins and machines" in Gale Crater . . . possibly, as "political insurance" against the day when (if this theory is correct) "*all* is to be politically revealed."

CREDIT: THE ENTERPRISE MISSION AND PETER VORIAS

Figure 34. Partial montage of diverse "Martian artifacts" *Curiosity* has imaged over the past two years.

In 2014, the Chinese joined this apparent "creeping disclosure" campaign, landing two robotic probes (called *Chang'e-3* and *Yutu* – "Jade Rabbit") on the Moon's Mare Imbrium . . . at "19.5 degrees"—proceeding over subsequent weeks to release astonishing lunar images *confirming* our decades-old discovery of "ancient, glass-like ruins on the Moon on official NASA photographs returned by the Apollo mission crews!"

CREDIT: CNSA

Figure 35. In December 2013, the Chinese National Space Agency (CNSA) successfully landed its first robotic mission on the Moon—named *Chang'e-3,* after the Chinese "Goddess of the Moon." This *Chang'e-3* lunar surface image represents stunning Chinese confirmation of what the Enterprise Mission initially discovered (on NASA Apollo data) in 1993—the existence of "ancient, glass ET cities on the Moon!" The striking "glittering glass towers" visible on this official Chinese image, extending along the horizon and upward into space . . . are the astonishing proof!

Finally, and most spectacularly, is ESA's current *Rosetta* mission to "67P" . . . which has revealed increasingly artificial-looking geometry in each successive imaging session . . . as *Rosetta* drifts ever closer to this clearly "*radically* anomalous object."

Including, in the latest 67P images released by ESA at this writing, features which look unmistakably like the remains of "gigantic, half-collapsed *buildings*"—with their exteriors long-since eroded almost totally away . . .

with their bent and mangled former "interior supporting girders" . . . blatantly exposed (below).

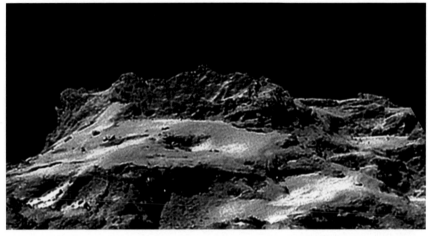

Figure 36. Enlarged portion of oblique *Rosetta* "Navcam" image, taken September 19, 2014—revealing another collapsed "mega-structure" on the limb of 67P . . . complete with exposed, girder-like supporting framework!

Three "independent," international space agencies . . . *one* likely now "creeping disclosure agenda". . . .

Will NASA's *New Horizons* Pluto mission successfully *complete* this "ancient ritual" . . . by, finally, *just telling us the Truth this July?!*

Stay tuned.

Addendum

As this chapter was going into "galleys," several major new developments in this unfolding drama suddenly took place.

The first: NASA's long-awaited announcement of the Hubble Space Telescope's triumphant discovery—to everyone's collective relief (after over twenty years of searching) in June 2014—of *three* potential "Kuiper Belt Objects for a *New Horizons* fly-by . . . *after* Pluto."[49]

The most likely of these—termed "PT1" (for "Possible Target One")—is estimated (from brightness assumptions only) to be only thirty miles across; if it was much brighter than the estimates of 3 or 4 percent, it could be *much* smaller.

PT1's orbit is so close to intersecting *New Horizon*'s current outward bound trajectory that only 35 percent of the *New Horizons* remaining on-board fuel will be required to effect a successful orbit change for *New Horizons* to make a direct intercept of PT1 in January 2019.

Four more years … and a billion added miles … *beyond* Pluto.

New Horizons represents, according to prevailing "mainstream planetary science," the first close-up observations of *a whole new class of solar system members*—termed loosely (by the IAU, in their controversial 2006 decision) "dwarf planets" (Pluto's current planetary status notwithstanding).

Long before the successful launch of the *New Horizons* Mission, one of the major goals for previously proposed "outer solar system mission to Pluto" (after the discovery of the first Kuiper Belt object, in 1992) had been "to also find a suitably-orbiting Kuiper Belt object for a *second* fly-by … after the primary Pluto Mission."

The Hubble discovery of PT1 last June insured (at least in the minds of a lot of intensely relieved planetary scientists) that *New Horizons* was going to answer at least some of the most fundamental mysteries remaining about the solar system's true origins.

We at Enterprise, looking at the same data from another point of view, had a very *different* set of "mysteries" in mind …

The second major space development that occurred just as we were wrapping up this chapter was that the European Space Agency (ESA) successfully landed (despite some serious technical anomalies) their *Philae* unmanned cometary probe on the surface of "67P" on November 12, 2014.[50]

Because of the (at this writing) still "undetermined anomaly" that occurred just at touchdown, *Philae* was unable to fire its two anchoring harpoons into 67P's surface. Thus, the lander did not become safely attached to the surface, and bounced right back off 67P into space, to a height of over *half a mile,* before gracefully (and oh, so slowly) falling back to 67P's extremely rugged surface, over the course of two whole hours.

According to the ESA analysis of "anomalous telemetry" received by *Rosetta* from *Philae* during this crucial and totally unplanned maneuver, the washing-machine-sized spacecraft also sailed (in its bouncing parabolic arc) more than half a mile *horizontally* from the original touchdown point, before coming to rest (after one more, "mini" bounce) in the shadow of a massive, looming cliff! This was much to the dismay of the heart-stopped *Philae* ground controllers sitting three hundred million miles away in Darmstadt, Germany, home of ESA's "European Space Operations Centre" (ESOC), which is controlling the *Rosetta* Mission.

Telemetry from *Philae*, processed by the ESOC computers, soon revealed that the too-near cliff was blocking "all but 90 minutes of sunlight each 'day'" out of 67P's full twelve-hour rotation period. Because of *Philae*'s angle to the sun, only a few watts of sunlight (at 67P's distance from the center of the solar system) were even reaching *Philae*'s solar panels in that brief time window—resulting in the probe being forced to conduct all of its eleven highly varied, pre-planned scientific investigations of the comet within the sixty-five-hour life of its non-rechargeable primary battery.

Pictured is an amazing time-lapse montage of images taken of *Philae*'s historic landing approach on 67P,[51] assembled by ESA from images shot over the space of about half an hour, from ten miles overhead, with the high-resolution OSIRIS camera aboard the *Rosetta* mothership.

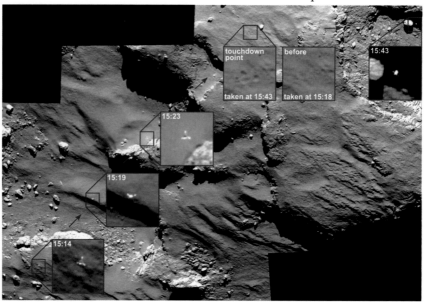

CREDIT: ESA AND THE ENTERPRISE MISSION

Figure 37. In this amazing time-lapse *Rosetta* montage—taken Nov 12, 2014 during Philae's historic descent to the surface of "Comet 67P/C-G"—we see a series of long-distance "snapshots" of *Philae* (bottom left to top right), inset over a mosaic of previous background comet surface images. Both sets were captured by the hi-res OSIRIS camera aboard the *Rosetta* "mothership," from a distance of just under ten miles. . . .

A comparison of "before" and "after" landing insets (top right), reveals the remarkable imprint of *Philae's* three, 16-foot-wide "tetrahedral landing pads" in the dusty "comet" surface ... before the spacecraft bounced over half a mile back into space, eventually coming to rest over half a mile away (last inset, far right) from the initial "landing/bounce point."

Amazingly, despite losing irreplaceable hours after landing trying to understand the nature (and the danger) of "the Big Bounce," *all* the pre-planned *Philae* scientific experiments, once initiated, were completed successfully before *Philae's* primary battery finally died and *Philae* was placed in electronic "standby mode."

The expectation was, based on calculations, that with the continued orbital movement of the comet around and closer to the sun over the next seven months, ever closer to "perihelion" (again, at this writing), it was predicted that *Philae* could, in fact, eventually store up enough solar energy to begin transmitting data—near the height of the expected "comet's" activity near perihelion. Perchance, just as Pluto is receiving its own first (modern) interplanetary visitor: *New Horizons.*

Both events will potentially take place dramatically—simultaneously—in July 2015.

If the images sent to Earth from the *Rosetta* orbiter have been astonishing—in terms of providing compelling scientific evidence of "ancient artificial structures on 67P"—the few *Philae* images transmitted *from the surface* have now added crucial "ground truth" to that previous evidence ... despite some additional "post-landing electronic anomalies" that inexplicably delayed the transmission of those first surface images from *Philae* by over twenty-four hours.

Pictured below is the first of two officially released ROLIS (Rosetta Lander Imaging System), 57-degree-wide descent images, taken as the *Philae* lander was slowly falling through 130 feet above the dusty surface.[52]

Figure 38. This is the first of two officially-released ROLIS (Rosetta Lander Imaging System) images, taken as the Philae Lander was slowly falling through the last ~130 feet above the dusty" comet" surface. The image is ~57 degrees (one radian) wide—to allow direct conversion of the "angular measure" of objects on the ground, into their absolute dimensions. Example: the size of the large "rock" (upper right)—calculated via Philae's altitude (130 feet) and the frame's "radian measure"—is about16 feet—the same as its observed width compared to the (known) 16-foot-wide span of Philae's fully-extended landing legs (Figure 37).

Also pictured is the same *Philae* image with added inset enlargements of several "highly anomalous *geometric* objects" appearing in the original 130-foot-wide-frame.

CREDIT: ESA AND THE ENTERPRISE MISSION

Figure 39. This is the same Philae image as in Figure 38 . . . with enlargement of several "anomalously geometric objects" highlighted by insets. Given that geology, on this scale, has no known processes capable creating such radically different geometries, in virtually zero gravity—and certainly in such close proximity—the more likely explanation is that these diverse geometric objects are exactly what they appear to be: pieces of abandoned, artificial "junk"—littering the surface of 67P from its epochs of former habitation. A detailed description of each potential artifact is included in the text.

The large, apparent "boulder" seen in the upper right (approximately a dozen feet across), on closer examination (inset) reveals a remarkable (if heavily eroded) "symmetrical, right-angle, three-dimensional *grid-pattern* on its upper surface"—reminiscent of some kind of "mechanical sub-systems" attached to its exposed surface, rising out of the pervasive surrounding dust, and looking like an exposed portion of a half-buried, ancient *machine*. The middle-left inset reveals a remarkable "bladed-looking" artifact (clearly, not a "rock") with ninety-degree edges and parallel, opposing sides; a curious, "faceted knob" is attached to the "blade" at one end.

147

Remarkably symmetric on one end, the long, flat object measures approximately fifteen feet long and four feet high—again, looking nothing like "a random rock," but something "carefully machined" (or, 3-D printed!).

The final *Philae* enlargement focuses on a tantalizing, almost invisible structure hidden in the upper left-hand corner of the original ROLIS frame: an extremely dark (even on a "cometary" body—with an average 3 to 4 percent reflectivity!) object, the feature appears vaguely menacing and not at all "natural." Multiple, much lower-resolution NavCam images confirm the reality of this dark structure, but shed no additional light on its function or purpose when it was new. The diagonal black bar crossing the upper right-hand corner of the image is a section of *Philae*'s own landing gear, seen in silhouette against the "comet's" surface. In the center of this image (above) is a small, bright, somewhat pointed feature (like a half-buried arrowhead) that becomes the focal point for the second, much closer ROLIS descent image (below).

CREDIT: ESA AND THE ENTERPRISE MISSION

Figure 40. In this second ROLIS image, taken just 30 feet above the surface of 67P, the largest object is a "pointed rock" about 5 feet across—with distinct, artificial-looking "flutes" seen prominently along one anomalously straight "double-edge." On enlargement (inset), the "flutes" resolve to a series of "square, regular indentations"—arranged parallel to the left-hand edge . . . connecting to another linear feature, also running parallel to that first edge. Such multiple, redundant symmetries and right angles—in such a small space—are all-but-impossible to achieve by any known natural process, leaving "artificiality" as the only logical alternative. To any objective observer, what "leaps out" of this ultra-close-up 67P surface image, is the amazing number of additional, blatant examples of clearly "eroded, mechanical junk" in this one small, random image. . . . Why are such multiple examples of clearly ancient, manufactured objects . . . even there!?

To any objective observer, what leaps out of this ultra-close-up 67P surface "aerial" image is the amazing number of blatantly *geometric* examples of "eroded, clearly mechanical junk" seen littering the "comet's" surface, laid out less than thirty feet below the still descending *Philae* spacecraft; the number of highly-angular, highly symmetric and regularly-incised, clearly *artificial* fragments of once much larger, manufactured objects visible just in this small area (about one hundred feet square) is astonishing, and utterly "impossible."

Unless "comet" 67P is, in fact, a long-abandoned, once inhabited derelict *spaceship* (!) now eternally adrift in interplanetary space between the planets Mars and Jupiter within the classic "asteroid belt."

The presence of even *one* clearly mechanical object in these *Rosetta* and *Philae* images would violate every "natural" assumption rigidly held dear by planetary scientists around the world; in this enlargement (below) of a portion of the ROLIS image #2, something shiny appears: looking for all the world like the orphaned head of a very familiar-looking, very "human-looking" artifact—a hammer!

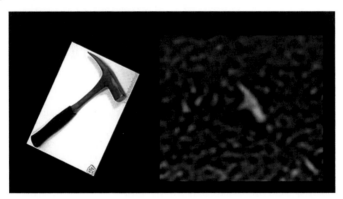

CREDIT: ESA AND THE ENTERPRISE MISSION

Figure 41. "If I had a hammer . . . ." The presence of even one verified, clearly mechanical object on 67P would violate every "natural" assumption currently rigidly held dear by planetary scientists around the world; in this enlargement of a portion of the ROLIS image #2, something longer than wide, symmetrical . . . and very "shiny," appears on 67P—looking, for all the world, like the orphaned head of a very familiar . . . very human-looking artifact on Earth—a hammer!

The odds of such a familiar, highly specific hammer-headed shape, randomly appearing on the surface of such a tiny, completely alien "mini-world" over three hundred million miles from Earth—along with all the other, equally "impossible" mechanical shapes sharing this same image—are simply beyond calculation, being (at least) "trillions to one" against!

Based on what is clearly visible in these two *Philae* surface frames, at what point, then, will the professional geologists at ESA, NASA, and CNSA, in looking at these two ROLIS descent images blatantly filled with "alien stuff," throw up their hands and grudgingly admit what is obvious to any honest layman: "All these familiar-looking things just *can't* be "geological!!"

But, beyond all the abundant *Rosetta* and *Philae* imaging evidence confirming that "67P is not an ordinary 'comet'—but, in fact, has all the appearance of 'an ancient, incredibly eroded, 'mega-habitat' in space," 67P had one more surprise to throw at us: a Signal.[53]

The third potential "tipping point event" occurred November 11, 2014: ESA's sudden announcement, just one day before *Philae*'s attempted landing on 67P, of *Rosetta*'s surprising detection of "a song"—coming *from* the "comet!"[54]

The most remarkable feature of this "song" (which is being created, in the mainstream model, by density variations in the electrified plasma—the coma—enveloping the "comet," creating resonant magnetic field variations in the process) was a very different . . . underlying . . . precisely clock-like frequency—driving the ESA recording.

If not caused by some obscure "cross-talk" issue in the magnetic instrument aboard the *Rosetta* spacecraft itself (in which case, one would expect that ESA would have explained that important caveat with their release of this new data), the remarkable *machine-like precision* of this underlying tone can *only* be due to some kind of relatively small, repetitive, *artificial* process operating somewhere on—or in—the "comet!"

Had the Europeans picked up (or, triggered with *Rosetta*) some kind of "torsion *homing* beacon," switched on by *Rosetta*'s close approach (within sixty miles) to 67P in early August 2014? The actual detection (and then public admission) by the Europeans of "some kind of signal emanating from 67P"—a signal having many of the proposed characteristics I had written about only weeks before, in this very book, re: the possibility of "detecting a deliberate Torsion Homing Beacon" at Pluto, detectable by *New Horizons* as a means of identifying "an ancient solar system Archive"—was indescribably exhilarating. Was *Rosetta*'s confirmation of this unique, long-wave RF emission from 67P, a sign that this extraordinary object was, in fact, not only an "ancient ET spaceship" but also *another* "ancient solar system Archive!?"

As the *Rosetta* mission has unfolded since last August, and the truly exotic nature of 67P has increasingly sunk in, celestial mechanics studies not only have discovered that 67P "yesterday" (in "astronomical reckoning")—in 1959—experienced a *major* orbit change (after a gravitational encounter with the massive Jupiter); that close passage "shrank" 67P's previous orbit, from a perihelion of 2.77 AU to a new (post-1959) orbit closest approach to the sun of 1.29 AU.[55]

As part of this inquiry into the long-term orbital history of Comet 67P/Churyumov-Gerasimenko, celestial mechanics experts "ran the clock even further back" before 1959 to a time *thousands* of years earlier, when, it was calculated, 67P likely had first been captured in the inner solar system (again, via the far-reaching gravitational effects of Jupiter) ... this time, from its much vaster, original orbit in the *outer* solar system—*in the Kuiper Belt.*

Was 67P then—as an *original* Kuiper Belt object—only a foreshadowing of the wonders still awaiting *New Horizons* and the rest of us, at Pluto . . . and beyond? Again, stay tuned.

Notes

1. *Wikipedia*, s.v. "Lost Horizon," last modified July 12, 2014, http://en.wikipedia.org/wiki/Lost_Horizon.
2. *Wikipedia*, s.v. "Shangri-La," last modified October 15, 2014, http://en.wikipedia.org/wiki/Shangri-La.
3. Richard C. Hoagland, "It Only Takes ONE White Crow: The Real Agenda of NASA's Curiosity Mars Rover Mission," *Enterprise Mission,* http://www.enterprisemission.com/Curiosity-White-Crow.htm.
4. Richard C. Hoagland, *The Monuments of Mars: A City on the Edge of Forever,* 5th ed. (Berkeley: North Atlantic Books, 2001), first published 1987.
5. Eugene Chiang and Gregory Laughlin, "The Minimum-Mass Extrasolar Nebula: In-Situ Formation of Close-In Super-Earths," *Monthly Notices of the Royal Astronomical Society,* May 12, 2013, http://arxiv.org/pdf/1211.1673v1.pdf.
6. Ibid.
7. Jonathan O'Callaghan, "Scientists Have NO Idea How Planets Form: Discovery of Hundreds of New Worlds Has Left Experts Baffled," *Daily Mail Online,* July 3, 2014, http://www.dailymail.co.uk/sciencetech/article-2679337/Scientists-NO-idea-planets-form-Discovery-hundreds-new-worlds-left-experts-baffled.html.
8. Ann Finkbeiner, "Astronomy: Planets in Chaos," *Nature: International Weekly Journal of Science,* July 2, 2014, http://www.nature.com/news/astronomy-planets-in-chaos-1.15480.

9. "Copernican Principle," https://www.princeton.edu/~achaney/tmve/wiki100k/docs/Copernican_principle.html.

10. Joe Garland, "The Kardashev Scale: Type I, II, III, IV & V Civilization," *From Quarks to Quasars,* July 19, 2014, http://www.fromquarkstoquasars.com/the-kardashev-scale-type-i-ii-iii-iv-v-civilization.

11. Hoagland, *The Monuments of Mars.*

12. *Wikipedia,* s.v. "Stargate SG-1," last modified October 3, 2014, http://en.wikipedia.org/wiki/Stargate_SG-1.

13. *Encyclopaedia Britannica,* s.v. "Clyde W. Tombaugh," last updated April 9, 2013, http://www.britannica.com/EBchecked/topic/598927/Clyde-W-Tombaugh.

14. *Encyclopaedia Britannica,* s.v. "Dwarf Planet," last updated August 22, 2013, http://www.britannica.com/EBchecked/topic/1224420/dwarf-planet.

15. Jennifer Rosenberg, "Pluto Discovered," *About.com,* N.D., http://history1900s.about.com/od/1930s/qt/Pluto.htm.

16. "Clyde W. Tombaugh," *New Mexico Museum of Space History,* N.D., http://www.nmspacemuseum.org/halloffame/detail.php?id=51.

17. "Transcript: The Girl Who Named Pluto: Interview with Venetia Burney Phair," *Nasa.gov,* January 17, 2006, http://www.nasa.gov/multimedia/podcasting/transcript_pluto_naming_podcast.html.

18. Todd E. Creason, "Walt Disney: Freemason or Not?," *Todd Creason* (blog), October 6, 2011, http://toddecreason.blogspot.com/2011/10/walt-disney-freemason-or-not.html.

19. Armand Diaz, "The Other Pluto: Getting to Know Walt Disney's Cartoon Dog," *Astrology News Service,* August 4, 2014, http://astrologynewsservice.com/opinion/the-other-pluto-getting-to-know-walt-disneys-cartoon-dog.

20. Clyde W. Tombaugh, "The Struggles to Find the Ninth Planet," rpt. April 4, 1995 on University of Arizona course website, http://ircamera.as.arizona.edu/NatSci102/NatSci102/text/ext9thplanet.htm.

21. John J. O'Connor, and Edmund F. Robertson, "Mathematical Discovery of Projects," University of St Andrews Scotland School of Mathematics and Statistics Archive, September 1996, http://www-history.mcs.st-and.ac.uk/HistTopics/Neptune_and_Pluto.html.

22. O'Connor and Robertson, "Ernest William Brown," University of St Andrews Scotland School of Mathematics and Statistics Archive, October 2003, http://www-history.mcs.st-and.ac.uk/Mathematicians/Brown.html.

23. http://www.hup.harvard.edu/catalog.php?isbn=9780674002913

24. Mike Brown, *How I Killed Pluto and Why It Had It Coming* (New York: Spiegel and Grau, 2012).

25. Slipher's biography can be read at http://www.roe.ac.uk/~jap/slipher.

26. "George Ellery Hale," *Mount Wilson Observatory,* N.D., http://www.mtwilson.edu/Hale.php.

27. Jaroslav Makarov, "First Human Buried in Deep Space: The Long, Cold Trip of Clyde Tombaugh," *New Space Age,* February 28, 2014, http://newspaceage.com/2014/02/28/first-human-buried-in-deep-space-the-long-cold-trip-of-clyde-tombaugh.

28. Jeff Foust, "Demote Pluto, or Demote 'Planet'?" *The Space Review,* August 28, 2006, http://www.thespacereview.com/article/692/1.

29. Benjamin Simon, "Pluto Is a Planet After All," *Inquisitor,* October 1, 2014, http://www.inquisitr.com/1513129/pluto-is-a-planet-after-all.

30. Alan Stern, "Unabashedly Onward to the Ninth Planet," *New Horizons,* September 6, 2006, http://pluto.jhuapl.edu/overview/piPerspectives/piPerspective_09_06_2006.php.

31. Malcolm Shaw, "The Mariner Mars Missions," Nasa.gov, last updated January 6, 2005, http://nssdc.gsfc.nasa.gov/planetary/mars/mariner.html.

32. Tony Phillips, "NASA's New Horizons Spacecraft Nears Its Destination," *Clarksville Online,* January 16, 2014, http://www.clarksvilleonline.com/2014/01/16/nasas-new-horizons-spacecraft-nears-destination-planet-pluto.

33. Richard C. Hoagland and Mike Bara, *Dark Mission: The Secret History of NASA,* enlarged and revised edition (Port Townsend, WA: Feral House, 2009).

34. *Wikipedia* s.v. "Sumerian Creation Myth," last modified September 16, 2014, http://en.wikipedia.org/wiki/Sumerian_creation_myth.

35. Hoagland, *The Monuments of Mars.*

36. Calvin J. Hamilton, "Pluto and Charon," 1997, http://www.iki.rssi.ru/solar/eng/pluto.htm#stats.

37. Robert P. Hoyt, "Obituary: Dr. Robert L. Forward," *SpaceRef,* September 21, 2002, http://www.spaceref.com/news/viewpr.html?pid=9328.

38. David S. F. Portree, "Pluto: Doorway to the Stars (1962)," *Wired,* June 14, 2014, http://www.wired.com/2014/06/pluto-doorway-to-the-stars-1962.

39. O'Connor and Robertson, "Mathematical Discovery."

40. Richard C. Hoagland, "A Most Hyperdimensional Eclipse … and Final Venus Transit," *Enterprise Mission,* http://www.enterprisemission.com/Hyperdimensional-Eclipse.htm.

41. http://www.allais.info/alltrans/nasareport.pdf

42. "NASA's Hubble Discovers Another Moon Around Pluto," *Nasa.gov*, July 20, 2011, http://www.nasa.gov/mission_pages/hubble/science/pluto-moon.html.

43. Hoagland, "Hyperdimensional Eclipse."

44. Christopher Loring Knowles, "At the Edge of 17," *Secret Sun* (blog), November 8, 2007, http://secretsun.blogspot.com/2007/11/at-edge-of-17.html.

45. Brian Stallard, "Rosetta's 'Rubber Ducky' Comet Compared to Los Angeles," *NatureWorld News,* August 19, 2014, http://www.natureworldnews.com/articles/8622/20140819/rosettas-rubber-dukey-comet-compared-los-angeles.htm.

46. Tom Van Flandern, "The Exploded Planet Hypothesis 2000," *Meta Research*, N.D., http://www.metaresearch.org/solar system/eph/eph2000.asp.

47. European Space Agency, "Space in Images," August 2014, http://www.esa.int/spaceinimages/Images/2014/08.

48. Jeff Quitney, "Voyager 2 at Neptune," YouTube video, posted November 6, 2012, https://www.youtube.com/watch?v=COI5LBpvDyU.

49. HubbleSite, "NASA's Hubble Telescope Finds Potential Kuiper Belt Targets for New Horizons Pluto Mission," October 2014, http://hubblesite.org/newscenter/archive/releases/2014/47/full.

50. Tushna Commissariat, "Rosetta Scientists Land Probe on Comet for First Time," November 2014, http://physicsworld.com/cws/article/news/2014/nov/12/rosetta-scientists-land-probe-on-comet-for-first-time

51. European Space Agency, "Osiris Spots *Philae* Drifting Across Comet," November 2014, www.esa.int/spaceinimages/Images/2014/11/OSIRIS_spots_Philae_drifting_across_the_comet.

52. European Space Agency, "Comet from 40 Metres," http://www.esa.int/spaceinimages/Images/2014/11/Comet_from_40_metres.

53. Miriam Kramer, "Listen to This: Comet's Eerie 'Song' Captured by Rosetta Spacecraft," November 2014, www.space.com/27737-comet-song-rosetta-spacecraft.html.

54. Space.com, "Rosetta's Comet Sounds Like This," www.space.com/27734-rosettas-comet-sounds-like-this-magnetic-oscillations-video.html.

55. European Space Agency, "Rosetta's Target: Comet 67P/Churyumov-Gerasimenko," http://sci.esa.int/jump.cfm?oid=14615.

4
.....

Pluto and the Death of God

J. F. MARTEL

Pluto appeared in the sky between the wars, as the Roaring Twenties gave way to the Great Depression and the world continued its demented march toward Auschwitz and Hiroshima. Those two catastrophes were still years away; but already in 1930, those who cared (or dared) to scry the forces at work behind the scenes could feel the rising heat.

The great writer D. H. Lawrence was among these seers. In the opening to a novel released two years before Pluto made headlines, he observed that it is precisely *because* we live in tragic times that we reject a tragic vision of existence: "The cataclysm has happened, we are among the ruins." A careful reader of Friedrich Nietzsche, Lawrence knew that the social and political troubles of his day were just the outward signs of a deeper crisis, one whose nature was as metaphysical as it was socio-historical. As the plot of *Lady Chatterley's Lover* illustrates, this crisis existed well beyond the theater of power politics: it touched the most intimate areas of human life. This was something no treaty, social program, or League of Nations could fix, a collective "loss of soul" operating at pandemic levels.

It was in the midst of this crisis that astronomers located the obscure object their calculations had assured them existed at the edge of our solar system. But before talking about Pluto, I want to delve a little deeper into the situation in which it emerged, so as to better understand its imaginal significance.

Nietzsche had put his finger on the nub of the problem two generations before Lawrence with his proclamation of the death of God. While at one level, the statement "God is dead" is squarely aimed at the Christian deity, in the context of Nietzsche's overall philosophy it has a much wider scope. "Christianity" is the name Nietzsche gives to what he considers to be the most sophisticated articulation of an ancient, deep-rooted cultural *asceticism*, endemic to the West. It is essentially a form of nihilism, entailing a metaphysical denial of the body-soul in favor of a transcendent principle in whose searing light our world appears as mere shadow. From its place beyond the confines of matter and psyche, the transcendent object—what Nietzsche calls "God"—becomes the basis for universal moral judgment in this world. It determines right and wrong, truth and falsehood; but since it is outside of space-time it can't communicate with us directly: a priest (or other expert) is required to judge the living and the dead in its name. This life-denying attitude derives from what Nietzsche called a "spirit of vengeance" that has driven Western culture since the days of Socrates. It is a force that sucks the lifeblood out of embodied being, effectively imposing on everyone an infinite spiritual debt that can never be repaid.

The most subtle and virulent form of asceticism for Nietzsche isn't the Christian faith per se but modern secularism, with its enshrinement of metaphysical *truth* as a thing of intrinsic moral value. That is why Nietzsche follows up the proclamation "God is dead" with the warning that even in death, the shadow of God "still looms" . . . and will continue to do so for a long time. The doctrine of truth—which persists in the humanities as well as the sciences—is the ghost of God, the pesky vestige of a principle whose unreality has already disclosed itself in the collective psyche.

Over and above its indictment of the old cosmic patriarch, "God is dead" is a mantra signaling the disappearance of *any* principle on which one might hang an absolute, the vanishing of the very basis for positing universal values of any kind. For Nietzsche the news is both good and bad. It is good news because, the rug of transcendence having being pulled from under our feet, it is only a matter of time before we are forced to adopt a different way of being. But it is also bad news, because the realization that transcendence is a fiction is so traumatic that we may well destroy ourselves sooner than embrace a new modality.

The way I understand it, the Nietzschean demotion of truth doesn't mean that truth is nonexistent. Nietzsche is talking about the end of truth as a monolithic value deserving unquestioned devotion in the manner of an idol. As the French thinker Gilles Deleuze explains in a beautiful treatise on Nietzschean philosophy, it isn't a question of whether truth exists but *whether all truths are equal*—and most importantly, *whether truth as such is necessarily good*. "There are imbecile thoughts, imbecile discourses, that are made up entirely of truths; but these truths are base, they are those of a base, heavy and leaden soul," he writes.[1] Consider the know-it-all who would dismiss a poetic description of the sunrise because "actually, the sun doesn't rise, it's the Earth that's turning," or the historicist who would repudiate the revolutionary by reminding her that "all revolutions turn out badly in the end." Both make use of base truths to undermine more profound ones. Their truths are "base" because they shrink the horizon of the possible, they presuppose a world that can be conquered by the intellect and that can be *judged,* as opposed to one that exceeds every ideation. The acolytes of truth want to banish all sense of a universe in which something *new* might arise at any moment to challenge our most basic suppositions as to the nature of the real.

"Illusions" such as sunrises, revolutions, and works of art can contain truths that, while having no basis in hardcore actuality, are more expansive and significant than any plain statement of fact. Indispensable though they may be from an empirical standpoint, facts speak only what has been the case *in the past*: they have no claim on the real potentialities of the now. So it isn't truth as such that Nietzsche said must come to an end so much as a "human, all too human" understanding of truth. Likewise, it wasn't God himself whose death Nietzsche proclaimed, but a similarly too-human conception of the divine.

The modern crisis encapsulated in the mantra "God is dead" is an epistemic one: it involves the way we think about the world rather than the world itself. The way Deleuze puts it in his interpretation of Nietzsche, modernity has replaced *knowledge* with *belief*. It is this switch from knowing to believing that has changed the situation. We can't claim to truly *know* anything anymore. Even the solidity of matter becomes suspect in the light of modern ideas of nature and the cosmos. The Enlightenment quest for truth has revealed uncertainty to be an inextricable attribute of human thought. The best we can do today is to *believe* we know. In this sense it is we, and not the Australian aborigines, who live in Dreamtime, rowing merrily down the stream of a life that's but a dream. The West has shattered its dearest illusions to uncover levels of the real that won't conform to its own metaphysical ideals. We are just taking our sweet time acknowledging it. Hence the crisis.

While in daily life we continue to act as though everything were OK, an abyss yawns beneath ordinary consciousness, a *groundlessness* the horror writer Thomas Ligotti called the "Shadow at the Bottom of the World."[2] If the old transcendent god continues to haunt us in a secular guise, demanding that we persist in holding absolute truths despite the evidence of our thoughts, it is because only through this belief can we trick ourselves out of facing the void. And so we fight the shadow of the abyss with the shadow of God. Here I return to the opening of *Lady Chatterley's Lover:*

> Ours is essentially a tragic age, so we refuse to take it tragically. The cataclysm has happened, we are among the ruins, we start to build up new little habitats, to have new little hopes. It is rather hard work: there is now no smooth road into the future: but we go round, or scramble over the obstacles. We've got to live, no matter how many skies have fallen.[3]

We've got to go on, because stopping would mean confronting the imponderable fact that we have been divested of every line of defense against the total what-the-fuckness of it all. The only absolute remaining behind the mirages and the chicanery of obsolete transcendence is absolute unknowing, absolute mystery.

Anyone who has had a bona fide panic attack knows what I mean by "groundlessness." The fear comes on when the egoic mind hits on the untruth of truth, or the truth of untruth. The world suddenly takes on a cold, menacing glare. It doesn't become meaningless; rather, it comes alive with meanings emanating from an inhuman presence beyond comprehension, an alien intelligence whose indifference to our plight gives it an aura of malignancy and cruelty. In 1927, three years before the Pluto discovery, Martin Heidegger's *Being and Time* proposed that anxiety wasn't a mental disturbance but the only *sane* reaction to the actual nature of reality. Anxiety, for Heidegger, is the "fundamental mood," the modern mindset *par excellence*—that which brings us closest to the essence of Being.

Modern humans are adrift in the unknowable, the most potent image of which may be the immeasurable vastness of outer space through which Planet Earth wafts like a wad of spit from the mouth of Cthulhu. It's true that there is no shortage of religions, philosophical models, esoteric traditions, and spiritual practices for those who seek to make sense of the situation (myself included). However, it is impossible to entertain any one of these models of belief without being aware of the fact that it exists as one "option" among many. Whether the one I pick is the most adequate is anyone's guess, especially when I realize that there are very smart people vouching for almost every position. We live in a metaphysically *flat* time, with all of these different models existing on the same theoretical plane.

Be that as it may, mystical experiences, magical encounters, extraplanar flights, synchronistic happenings, and prophetic visions still abound. That is heartening because it tells us that secular materialism is really just one more option on the table, not the sober default its proponents would like it to be. On the other hand, even the most miraculous occurrence cannot restore the ground we have lost. A dozen mystics will return from the otherworld with a dozen different maps of meaning. And while we can think our way to a synthesis that makes, say, G. I. Gurdjieff's cosmology compatible with Rudolph Steiner's, the critical differences in need of bridging are a case in point. It isn't that the mystics are wrong; evidently, Gurdjieff and Steiner saw something—something important—in their visions. But in turning this something into a set of communicable ideas, they altered it in a significant way. The figurations of the real described by Gurdjieff, Steiner, and others seem to have as much to do with who did the seeing as they do with what was seen—and that is *modernity* defined.

As Carl Jung observes in the *Red Book,* if a new divinity, a new ground, should arise, it must inhere in the *relative,* that is to say the mysterious, the innately ambivalent: *both/and* as opposed to *either/or.* The pluralism that characterizes the spiritual scene today is not going away. We must somehow find our footing in groundlessness itself.

The discovery of Pluto coincides with the modern revelation of ground-lessness. Synchronistically speaking, Pluto is the emissary of the death of God, expressing it through the death god's return to mass consciousness.

It is worth noting that Pluto is not equivalent to the Christian devil (that dubious honor goes to Pan). In fact Pluto hardly appears at all in the Christian mythos, with good reason. For what room is there for the personification of nothingness in a cosmos where nothingness was done away with on the First Day? If there is one thing that traditional Christianity did splendidly, it was to reassure people that death was not a void, that the personality survived it intact. (It is true that it still does this for millions of Christians today. But again, modern Christians *believe* their creed with the full knowledge that it is one possibility among many. Far from being self-evident, contemporary Christianity must vie with its competitors. It therefore contains *doubt* as a constitutive element.) In the traditional Christian worldview, death isn't a real event but an optical illusion to be dispelled on the Day of Judgment. The fact that the body expires is completely irrelevant, because time and space themselves are irrelevant in the Kingdom of God.

It was the decline of Christianity as the preeminent Western mindset that allowed for the reemergence of Death as an archetype, and Pluto must be seen in the light of this epochal return. After Pluto, death once again becomes the ultimate horizon it had been for the people of antiquity. During the Cold War, its presence took the form of the atomic bomb (remember the key role of plutonium in making nuclear weapons), while today it looms above us still in the guise of global climate change (an eventuality that evokes a singularly Plutonian image of the Earth as a dead planet). Pluto can—should, I think—be interpreted as a symbol of the finality of death in a cosmos deprived of the metaphysical axis required to give existence an intelligible teleology.

We have been living under the sign of Pluto now for nearly a century, which may explain part of our fascination with doomsday scenarios. Nor does the fact that Pluto may have lost its official status as a planet indicate a change in the situation. It only makes sense that a planet named Pluto would turn out to be a dubious planet in the end: its namesake was the god who was not a god. In the ancient world Pluto was a deity without a temple, shrine, or cult, an invisible power dwelling apart from his Olympian siblings in the dank and lonely underworld. As James Hillman said, Pluto is the entity who has no substantial existence and yet in whose presence all things become manifest.[4] Whether as planet, planetoid, or "swarm," to borrow Richard Grossinger's term, Pluto remains what it is: the outer limit of our groundless world. Our black horizon.

Since I started thinking about this, it has occurred to me that I am able to summon up a mental image of every major Roman god except Pluto. When I try, I get the image of Mickey Mouse's pet dog instead. Incidentally, Pluto the Dog was born the same year Pluto the God reincarnated as Planet X (though it took until 1931 for him to get his name). Is it going too far to see a connection here? Think about it: just one year after a previously unknown object is named Pluto on account of its darkness, invisibility, and ambiance of futility, a ridiculous cartoon character bearing the same name enters popular consciousness with enough punch to become the first thing that pops into most heads when they hear the P-word. Silly as it seems, my sense is that Disney gave us Pluto the Dog as a distraction from the revelations of Pluto the God. Just as the three-headed dog Cerberus guards the doors of Hades, so Disney's bumbling mutt (and the whole entertainment culture he stands for) protects our day-world from the groundless night whose sign was Pluto.

Disney's scheme to defeat a god with a dog was by all appearances unconscious. In all probability, Walt just liked the sound of the word. But that doesn't necessarily raise doubts about the scheme's intentionality. By their fruits ye shall know them: ever since the wars we have been bombarded by cartoony silliness, the purpose of which is to distract us from the increasingly dire realities hemming us in, be they political, economic, or ecological. There is no "they" doing it; we are doing it to ourselves, because we do not want to see that the ground is gone, that, as Alan Moore puts it, "the world is rudderless."

If this deployment of Pluto the Dog—and of the incessant fantasia that followed him—was unconscious, it should come as no surprise that it revealed more than it would have if it had been a deliberate tactic. In his very first appearance in a 1930 animated film, Pluto (then named Rover) accompanies Mickey and Minnie on a roadside picnic. While the talking mice dance around the meadow in blissful ignorance, wild animals proceed to devour every scrap of food they have brought, proving Pluto to be rather sub-par as a guard dog. Then a storm erupts, putting a soggy end to the proceedings, and Mickey finally finds a use for the dog by sitting him on the windshield and making him use his tail as a wiper for the rain.

Even in this first attempt to upstage the death god, the ultimate futility of the enterprise betrays itself. Pluto the Dog is no match for Cerberus, the indifferent monstrosity of nature. Eighty-five years after the film's release, the storm has indeed broken over the picnic that is consumer society. Spectacle is becoming less effective at distracting us from the unfolding disaster. Like Mickey and Minnie, we are stranded in the rain with our cartoon dog. The hubris of Enlightenment had us strutting for a while; but today our invention and cleverness are turning against us, and we must struggle to reverse the processes we ourselves have set in motion. The future has never looked quite as Plutonian as it does now. Within a few centuries' time, Earth itself may look as deserving of the name Pluto as any other lifeless space rock.

We cannot go on imagining that we are the center of the universe or that we have God's number. Pluto's lesson, I think, is the same as it was when it appeared on the astrological horizon: a lesson of humility. We have been postponing it for almost a hundred years. If humanity is to survive the transition from transcendence to immanence foretold by Nietzsche, we must learn to fall to our knees again—only this time, it can't be before our own image. We must learn to bow before the mystery itself, and embrace it as the basis of a new religiosity.

Pluto's reappearance in the modern era calls us back to a time when the death god was acknowledged, however quietly, as an essential dimension of the phenomenal world. Perhaps it is in the nature of this god to reveal that all of our perambulations have been circular, and that we stand today in the same position as the ancient Greeks did when the tragedy of existence brought them face to face with him. Even then, people wanted to turn away and remain blind, like the characters in *Lady Chatterley's Lover,* to the tragic nature of a tragic situation. But then as now there were exceptions. Called "the Obscure" by his contemporaries, the fifth century BC philosopher Heraclitus spoke in a language no one understood. Even the tone of his writings suggests he was writing for the people of another time. My sense is that Heraclitus was writing for us now, albeit even in 2015 he is perhaps best described as a philosopher of the future. Heraclitus is one of the few thinkers to have the courage to seize on *becoming* as the deepest stratum of Being. In his philosophy, Being has no fixity: what is *is* insofar as it changes, and changes constantly. Groundlessness was Heraclitus' rule: he was able to look out at the world without projecting a transcendental axis to stabilize the vision into a tolerable dream. Flux, for him, was the only truth . . . the truth of untruth. Even the most sophisticated ideas or concepts could not in his eyes do justice to the whole.

Heraclitus famously remarked that the Eternal was not a patriarch like Yahweh or Zeus, but a "child at play." What he called the "Logos" was the form of change itself, the pure difference inhering in all things, endlessly incarnating itself through acts of creation and experimentation. Long before Nietzsche, Heraclitus knew that God was dead, but this didn't stop him from speaking of God as Logos. This isn't a moralistic God, but it is not an immoral one either: the Logos is *transmoral,* encompassing the reality of good and evil, yet subsuming them in something bigger. Nor is it useful to argue over whether this force of Becoming belongs to order or chaos: it is both/and, what James Joyce called a "chaosmos."

Heraclitus can serve as the beginning of a spirituality of immanence that allows meaning and purpose to survive, in new forms, in a post-transcendent world. In him, we find a vision that philosophy turned away from just as it was getting started, but that artists have embraced, often unconsciously, since the dawn of time. It took two and a half thousand years of transcendental idealizing to reach the point where the fiction won't stick anymore. Circumstances have made it such that, terrible or not, the pre-Socratic Heraclitean vision of eternal flux imposes itself on all of us. The survival of our species seems to hinge on whether or not we are able to accept it.

Throughout his career, Carl Jung implored us to turn to our own personal darkness, recognize our shadows, and stop projecting them onto other people and institutions. "We are the origin of all coming evil," he said. The global catastrophe that is upon us now is a direct outcome of our refusal to face the horror. When the great anthropologist Knud Rasmussen asked the Inuit shaman Aua what his people believed, the latter replied: "We do not believe. We fear." Meaning: we see that no belief can encompass the whole, and that it is precisely in hypostatizing metaphysical models of the totality that we blind ourselves to Becoming and expose ourselves to danger. Aua knew that it takes tremendous courage to fear consciously. Traditional Inuit culture has no God—no univocal, monolithic signifier. Nevertheless it has thrived in the harshest climate on Earth for millennia and enabled some of the most penetrating explorations of the imaginal realm that have ever been undertaken, in terms of both shamanic practice and artistic creation. I mention this because there are some who, upon hearing the phrase "God is dead," think that it means the end of spirituality because in God's absence, the dimension of the sacred evaporates. In fact, I think it's the opposite: only once God is dead (or, what is more exact from a Plutonian perspective, only once *God is death*) do the most expansive visions of the divine become possible. Here we stand not only with the shamans and artists, but also with the great mystics of the organized religions, all of whom experienced, in solitude, the atheistic crisis that we are about to undergo as a civilization.

In some esoteric passages Deleuze speaks of the importance of "affirming chance." I interpret this to mean a few things. First, it means accepting that everything the egoic subject wishes were the case (destiny, truth, divine justice, deliverance, even enlightenment) are only *relatively* true. In giving them an absolute status we effectively construct an image of the world that must be reaffirmed against all the vicissitudes of reality, which contradict it at every turn. Second, to affirm chance means to recognize that vaporizing our fictions ("killing the Buddha," as they say in Zen) can be as liberating as it is terrifying. The ego gives in to a deeper selfhood that, though it is no longer reducible to a personality, opens up new possibilities. Indeed, through the affirmation of chance literally everything becomes possible, since the divine judgment that limited the scope of the real is finally consigned to silence. Third, affirming chance means recognizing that the mystery of being is one that no religious gnosis, philosophical system, or scientific discovery will ever solve. Existence inheres in mystery, not in the sense that the riddle of life has a final answer we don't *yet* know, but in the sense that *knowing*— actually getting to the bottom of anything—*is simply something we humans don't do.* Truth ceases to be a function of power by which one might pass judgments about what is real or unreal, possible or impossible, only when we recognize that *mystery is constitutive of reality as such.* It is through this interruption in the closed circuit of the cosmos that the holy gets in. It is in groundlessness that the sacred comes to permeate every particle of the world.

In one fragment, Heraclitus states that Pluto and Dionysus are one and the same.[5] At first glance, no two gods could seem more opposed to one another than these two. Pluto is the lugubrious king of silence and death, Dionysus the wild prince of music and life. But on closer inspection the equation bears out. Dionysus is the god who descends into Hades in order to rise again as the "twice-born." He is the one who, by accepting death as a dimension of life, is able to attain to the plenitude. In Nietzsche's philosophy, Dionysus embodies all that is "active" in the universe, all those creative forces that precede and overcome the "passive" passions of the reactive self, which for their part secrete only nihilism and judgment.[6] The active forces encompass those aspects of life that the instruments of reason cannot measure: magic, miracles, synchronicity, vision, creation, self-giving . . . Though they are as present today as they always have been, the active forces can now only manifest in the private experience of isolated individuals or small groups of people—their reality cannot be shared at the societal level. I think the time is coming when it will be necessary for our civilization to see them once more as an integral part of the universe.

The apocalypse announced by Pluto in 1930 is now upon us. Night is falling. If we continue to resist, then we will almost certainly descend into a true and final darkness. But perhaps there is still a chance to come to terms with the shadow at the bottom of the world, to tap into its dark energy in order to transform our world and ourselves into some new form of being. Could that be what the aptly-named Grateful Dead were getting at with those obscure lines from their Plutonian ode, "Dark Star": *"Shall we go, you and I while we can, / Through the transitive nightfall of diamonds?"*[7]

Notes

1. Gilles Deleuze, *Nietzsche and Philosophy,* trans. Hugh Tomlinson (New York: Columbia University Press, 2006), 105.
2. Thomas Ligotti, *Grimscribe: His Life and Works* (Burton, MI: Subterranean Press, 2011), 215–226. "The Shadow at the Bottom of the World" is the title of the collection's closing story, which depicts a small rural community's hopeless confrontation with the abyss.
3. D. H. Lawrence, *Lady Chatterley's Lover* (London: Penguin Books, 1960), 5.
4. James Hillman, *The Dream and the Underworld* (New York: Harper & Row, 1979), 27–32.
5. Fragment 15. Heraclitus uses the name Hades for the death god.
6. See Deleuze, *Nietzsche and Philosophy*, for more on active and passive forces in Nietzsche's thought.
7. Robert Hunter and Jerry Garcia, "Dark Star" (song lyrics), accessed April 22, 2014, http://www.allmusic.com/song/dark-star-mt0010693242/lyrics.

5

Hades

JAMES HILLMAN

Hades was of course the God of depths, the god of invisibles. He is himself invisible, which could imply that the invisible connection is Hades, and that the essential "what" that holds things in their form is the secret of their death. And if, as Heraclitus said, Nature loves to hide, then nature loves Hades.

Hades is said to have had no temples or altars in the upper world and his confrontation with it is experienced as a violence, a violation (Persephone's rape; the assaults on simple vegetative nymphs, Leuce and Minthe; and *Iliad* 5, 395 or Pindar *Ol* 9, 33). He is so invisible in fact that the entire collection of Greek antique art shows no ideal portrait of Hades, such as we are familiar with of other gods.

He had no representative attributes, except an eagle, which brings out his shadowy affiliation with his brother, Zeus. He leaves no trace on Earth, for no clan descends from him, no generations.

Hades's name was rarely used. At times he was referred to as "the unseen one," more often as Pluto ("wealth," "riches") or as Trophonios ("nourishing"). These disguises of Hades have been taken by some interpreters to be euphemisms for the fear of death, but then why this particular euphemism and not some other? Perhaps Pluto is a description of Hades, much as Plato understood this god. Then, Pluto refers to the hidden wealth or the riches of the invisible. Hence, we can understand one reason why there was no cult and no sacrifice to him—Hades was the wealthy one, the giver of nourishment to the soul. Sometimes, he was fused with Thanatos ("Death") of whom Aeschylus wrote, "Death is the only God who loves not gifts and cares not for sacrifices or libation, who has no altars and receives no hymns . . ." (frg. *Niobe*). On vase paintings when Hades is shown, he may have his face averted, as if he were not even characterized by a specific physiognomy. All this "negative" evidence does coalesce to form a definite image of a void, an interiority or depth that is unknown but nameable, there and felt even if not seen. Hades is not an absence, but a hidden presence—even an invisible fullness.

Etymological investigations into the root word for *death demon* show it to mean "hider."

To grasp better the ways in which Hades hides invisibly in things, let us take apart this concept, listening for the hidden connections, the metaphors, within the word *hidden* itself: (1) buried, shrouded, concealed from eyesight, whether a corpse or a *mysterium;* (2) occult, esoteric, concealed in the sense of secret; (3) that which per se cannot be seen: non-visible as non-spatial, non-extended; (4) without light: dark, black; (5) that which cannot be seen on inspection, i.e., blocked, censored, forbidden, or obscured; (6) hidden, as contained within (interior) or as contained below (inferior), where the Latin *cella* ("subterranean storeroom") is cognate with the Old Irish *cuile* ("cellar") and *cel* ("death"), again cognate with our hell; (7) that which is experienced with dread and terror, a void, a nothing; (8) that which is experienced as hiding, e.g., withdrawing, turning away from life; (9) stealth, surreptitiousness, deceit, such as the hidden motives and unseen connections of Hermes. In short, Hades, the hidden hider, presides over both the crypt and cryptic, which gives to Heraclitus' phrase (frg. 123): "Nature loves to hide" *(physis kryptesthai philei)* a subtle and multiple implication indeed.

Some say that the cap or helmet Hades wears belongs primarily to Hermes and may have little or nothing to do with Hades.

This hat is a curious phenomenon: Hermes wears it, Hades wears it; Athene puts it on to beat Ares, and Perseus to overcome the Gorgon. It makes its wearer invisible. Evidently the explicit image of connection between Hermes and Hades (announced in the Homeric "Hymn to Hermes") is the headdress. Hermes and Hades share a certain style of covering their heads that both hides their thoughts and perceives hidden thoughts. It is their intentions that become invisible. We cannot perceive where their "heads are at," though we may have the sense of a hidden watch over our inmost thoughts. Because we can never discover what their covert minds intend, we consider them deceptive, unpredictable, frightening—or wise.

When we consider the House of Hades, we must remember that the myths—and Freud too—tell us that there is no time in the underworld. There is no decay, no progress, no change of any sort. Because time has nothing to do with the underworld, we may not conceive the underworld as "after" life, except as the afterthoughts within life. The House of Hades is a psychological realm now, not an eschatological realm later. It is not a far-off place of judgment over our actions but provides that place of judging now, and within, the inhibiting reflection interior to our actions.

This simultaneity of the underworld with the daily world is imaged by Hades coinciding indistinguishably with Zeus, or identical with Zeus *chthonios*. The brotherhood of Zeus and Hades says that upper and lower worlds are the same; only the perspectives differ. There is only one and the same universe, coexistent and synchronous, but one brother's view sees it from above and through the light, the other from below and into its darkness. Hades's realm is contiguous with life, touching it at all points, just below it, its shadow brother *(Doppelgänger)* giving to life its depth and its psyche.

Because his realm was conceived as the final end of each soul, Hades is the final cause, the purpose, the very *telos* of every soul and every soul process. If so, then *all* psychic events have a Hades aspect, and not merely the sadistic or destructive events that Freud attributed to Thanatos. All soul processes, everything in the psyche, moves toward Hades. As the *finis* is Hades, so the *telos* is Hades. Everything would become deeper, moving from the visible connections to the invisible ones, dying out of life. When we search for the most revelatory meaning in an experience, we get it most starkly by letting it go to Hades, asking what has this to do with "my" death. Then essence stands out.

Here too Hades has bearing on psychological theory. A psychology that emphasizes the final point of view—Jung's, for instance, and Adler's—is restating the Hades perspective, even if these psychologies do not go right to the end of their ends. I mean by this that the finalism in psychology seems to shy away from the full consequences of mythology, in which the finalism is not a theory only but is the experience in soul of its call to Hades.

Now hold here a moment. Let us beware of taking this call as literal death, of which so much is spoken and written today, that we begin to believe we know all about what we know nothing about. Literal death is becoming a clichéd mystery, that is, we have best-seller evidence about the unknowable.

Rather, by the call to Hades I am referring to the sense of purpose that enters whenever we talk about soul. What does it want? What is it trying to say (in this dream, this symptom, experience, problem)? Where is my fate or individuation process going? If we stare these questions in the face, of course we know where our individuation process is going—to death. This unknowable goal is the one absolutely sure event of the human condition. Hades is the unseen one and yet absolutely present.

The call to Hades suggests that all aspects of the process of the soul must be read finally, not only as part of the general human process toward death, but as particular events of and in that death. Each facet then is a finished image in itself, completing its purpose that is at the same time unending, not literally unending in time but limitless in depth. In other words, we can stop nowhere—and anywhere—because the end is not in time but in death, where death means the *telos* or fulfillment of anything, or, we can stop anywhere, because from the final point of view everything is an end in itself. The goal is always now.

A true finalistic psychology will show its ends in its means. We will be able to see its end goal of death in the methods it uses to work toward it. Therefore, to live fully into the consequences of the finalistic view means to bear the perspective of Hades and the underworld toward each psychic event. We ask: what is the purpose of this event for my soul, for my death? Such questions extend the dimension of depth without limit, and again psychology is pushed by Hades into an imperialism of soul, reflecting the imperialisms of his kingdom and the radical dominion of death.

Notes

1. M. P. Nilsson, *Gechichte der griechischen Religion,* 2nd ed. (Munich, 1955), 1: p. 452.

2. "The Misfortunes of Mint," ch. 4 of Marcel Detienne, *The Gardens of Adonis* (London: Harvester Press, 1977).

3. W. H. Roscher, *Ausführliches Lexikon der griechischen und römischen Mythologie.* "Hades," reprint, Hildesheim: Olms, 1965.

4. L. R. Farnell, *The Cults of the Greek States* (New Rochelle, New York: Caratzas, 1977), 3: p. 286.

5. Ibid., p. 283.

6. Augustus Pauly, ed. rev. by Georg von Wissows *et alia, Realencyclopädie der klassischen Altertunswissenschaft* Stuttgart: Metzger and Alfred Druckenmüller, 1893), "Thanatos."

7. Farnell, op., cit, p. 287.

8. E. Herzog, *Psyche and Death,* trans. D. Cox and E. Rolfe (London: Hodder and Stoughton, 1966), p. 39 f. on etymology of various death demons. For full discussion of the etymology of Hades, see Hjalmar Frisk, *Griechisches Etymologisches Wörterbuch* (Heidelberg: Carl Winter, 1973), 1: pp. 33–34. Plato is an important authority for the etymological identification of Hades with the "unseen" or "invisible (*Gorgias* 439b, *Phaedo* 80d, 81c). L. *Wächter,* "Zur Ableitung von 'Hades' und 'Persephone." *Zeitschrift f. Relig.-u. Geistesgeschichte* (1964): 193-97, derives *Hades* not from "invisible," but via Semitic roots from the underworld "waters" that rise from the deep. This derivation does not alter the equation: underworld = unconscious or id; for Freud often spoke of the id as a "reservoir," just as Jungians speak of the unconscious as a "source" and as "coming up."

9. H. Lloyd-Jones, "A Problem in the Tebtunis *Inachus*–Fragment," *Class. Rev.,* n.s., vol. 15, no. 3 (1965): pp. 241–43.

6

Excerpts from *Pluto*

FRITZ BRUNHÜBNER

Editor's note: This is a reprint of material from 1930, an early take on the astrology of a new planet soon after its discovery.

Pluto (Latin: *Pluto*, the rich one) is the Zeus of the lower regions and the depths of the earth. He was ruler over the deceased also. The character of this rulership was the darkness and the formless invisibility (shadows). His dwelling was in the depths and, therefore, "domos Aides," he himself was Aides or Aidoneus, i.e., the invisible. An old symbol of this invisibility is the so-called helmet or Hood des Aides, which conforms to the hiding cape in Germanic mythology. The hood is made from the scalp of a dog. The helmet producing invisibility is the symbol of the rulership of Pluto over the invisible. After the fall of his father, Kronos (Saturn), he received the nether-worlds, at the time of the partition of the world, where he rules at the side of Persephone (Latin: *Proserpina*), whom he had kidnapped. He was the son of Kronos and Rhea, and brother of Zeus and Poseidon.

Pluto, who was the antithesis of the God of the Sun, Apollo, at first passed as the irreconcilable enemy of all new life, into which he continually sent death and destruction. But there is another, milder version of this myth:

The people also knew of this God of Death and the lower world, but more from life in nature. They were much better acquainted with the image of a creative active god living inside of the earth, and similar to the Italianic God, Tollumo, a god of fecundation, Pluto, therefore, passed as an active god inside the earth, a god of the bounties of the earth and distributor of riches. In addition to believing him as connected with the bounties of grain, he was also thought of as related to prosperity in mining. He was also known by other names, such as Host, Hunter, Zeus of the deceased, assembler of the people, the equal distributor (judge, administrator of Justice). Sometime Hades-Pluto was called "the door-keeper" and he was pictured with a key in his hand.

The whole surrounding of Pluto was thought to be very sad and gloomy. People pictured the place of the underworld as a region rich in volcanic nature and gloomy appearing natural phenomena, with cave-like passages, hot springs and mephitic effluvia. Sacred to Pluto was the narcissus and the cypress, as well as the *adiantum* (maidenhair). Black sheep were sacrificed to him with averted faces. . . . .

As to the fundamental nature of Pluto, he is a mixture of water and fire. Contrasts prevail here, but the contrasts are such as to combine and supplement. Water extinguishes fire, while fire turns water into vapor. Pluto is the planet of self-rediscovery, or the union of these two elements. However, the water principle is the stronger. Thus, we recognize once again the close connection between Pluto and the zodiacal sign Scorpio.

Pluto has a kinship with Jupiter in that he is multiform and gives abundance with extreme goals. Yet from these extremes an amalgamation results. On account of its enormous driving force, Pluto tries to seek new connections again and again.

If we desire to grasp the characteristics of Pluto correctly, we must first clearly understand that Pluto is a polarity, a "double-boundedness." Pluto has a double face—a Janus face—and we must distinguish between two evolutions, i.e., a lower and a higher vibration, similar to those of the sign Scorpio. . . .

Pluto kills or destroys but builds out of the elements of the destroyed— out of the old, something new; Pluto rises from the ashes like the phoenix. Pluto is transformations—transformations of one force into another force, of one element into another; it is transmutation and the *"alchemistra redivivius."*

Pluto is connection, transition, passage, bridge, boundary; the end and at the same time the beginning; it unbinds and binds; it brings revolutionary upheavals; it is the turning point. Pluto leads from torpidity to revival; from one condition—one condition of consciousness—into another, from one being into another being, from this life into the life hereafter. For this reason Pluto has been called in mythology the lord of the realm of the dead, the realm between, the astral world. Pluto is the overthrow of the old, sensing the new: the end of the old world and the ascent of a new spiritual epoch.

It is therefore not a mere chance that Pluto was discovered on the borderline of two ages, the turning point of human evolution. If Uranus is the first stage of the coming Aquarian age, then Pluto is the second stage. Pluto leads out of death, rigidity, a cramped state, through the stage of fermentation, preparation and development to revival, enlightenment, elevation, and clarification. Pluto will lead humanity out of the mechanization and mechanical technology of our times into an epoch of revivification, resurrection, of magic and creative power.

The Plutonic force acts in a renewing, enlivening, reviving manner, bringing forth the new; it is stimulating, preparatory, enthusiastic, breaking forth, breaking open, arousing, germinating, up-stirring, up-rooting, eruptive, up-raising, revealing, reorganizing, revolutionary. However, it is not revolutionary in the sense that Uranus is. Pluto brings only that which has been developed under cover, in secret (underground) into daylight, when the time is ripe. The force of Pluto is like wine in the stage of fermentation, which is working until it has gained access to air, or like a volcano that bubbles and seethes inside until it erupts with elementary force.

Seething, boring, hatching, brooding, evaporating, decomposing, fermenting, transforming; disunion, germination, procreation—all these are Pluto characteristics. All chemical and transforming processes , distillation, chemistry, spagyric, alchemy, metamorphosis—all belong to Pluto. . . .

Pluto is an exalted yet sublimated scorpion-like Mars, and is in the first place the purest, highest, potentialized mental energy, or creative willpower, similar to that possessed by the "magic man" (fakir, yogi, magician, adept, artist, genius).

The following is a preliminary compilation:

(a) On the material plane: Death, destruction, decomposition, struggle, force, fanaticism, adventurous desires, temptation, defiance, envy, intensification, sensuality.

(b) On the spiritual plane: Regeneration, reorganization, transformation, transmutation, creator of new life-giving force, form impulse; gives energy to strive for spiritual power and leadership, conscious clairvoyance, magic, mystic, magnetism, exaltation. . . .

All large earth movements, eruptions, geologic displacements, earthquakes covering a larger range, are under the influence of Pluto. Uranus causes the sudden shocks of a more local nature. Pluto is in many respects an intensification of the Uranian influence, especially in regard to such conditions that are released through impact and energy. So in this respect, Uranus and Pluto supplement, intensify, and vie, so to speak, with each other. However, the effect of Pluto is never so surprising and sudden as in the case of Uranus.

Pluto conceals within himself a very important physical law: the ability to overcome the force of gravity and air pressure. In this category belong all the bold and adventurous technical problems involved in stratosphere flight, rocket flight, and the exploration of the great ocean deeps. . . .

Just like the other slow-moving planets, Pluto defines more favorable or unfavorable periods, rather than certain dates. The influence of the Pluto transit cannot be fixed to a certain day. It is possible, however, to narrow down the time of the release if one observes which directions or similar transits are due at the period influence of the transit of Pluto. My present experience indicates that one does not get too far by allowing three to four years for the periods of influence for the Pluto transit. Pluto's influence is, however, not always at its strongest when the aspect is exact. The effect of Pluto, above all, is felt, similar to Saturn, when the aspect is separating and only seldom in its application; the majority of the aspects were released 1 to 2 degrees after the exact point, as far as I have been able thus far to ascertain. . . .

According to my observations, the strongest effects have been the transits of Pluto over the cusps of the angles, over the Sun, Moon, and the superior planets, especially of his neighbors, Neptune and Uranus, and relatively, the weakest are the transits over Mercury and Venus.

Transits over the ascendant bring revolutionizing tendencies, a turnabout in development, an awakening of latent powers, turning points in the course of life, and the beginning of an entirely new phase in destiny. This is a time of decisions, temptation, trials, an enormous test of power, self-examination, and a standing at the parting of the ways. For many, it is the baptism with fire; of bringing forth the last urge one can master, creating intensified impulses and instincts. The nature wrestles with deep problems.

The favorable transits, and also the conjunction in the few cases where one can classify it as favorable (for instance), bring the tendency to strengthen the physical powers, and give record performances, both bodily and spiritually; of being entrusted with a special mission; an increase in the native's personal influence; an unexpected rise; an impetus toward experience that shapes the future; a raised receptivity toward astral influences or inspiration.

The unfavorable transits are outspoken danger signals and as a rule bring enormous exertion of the native's personal strength; strong emotional upsets; grave illnesses, and danger of accidents. There may be diseases of the blood, heavy blows of fate which often occur in series; sacrifice; renunciation; death of very near persons, also dangers to the native's own life. If other similar configurations occur, death may come. . . .

Uranus and Neptune can be classified as the forerunners of Pluto. Uranus, which was the first in the row of trans-Saturnian planets, acted as an evolutionizing, revolutionizing force; he brought inventions and set technic upon the throne. Uranus also represents the first knowledge of transcendental matters, and the first bridge between matter and spirit. The influence of Uranus is not yet purely spiritual, but more material-spiritual.

As a continuance of the Uranian influence and his first awakening of a new epoch in time, Neptune at first brought confusion and chaos into human experience because his influence was not understood. But later, as mankind tried more and more to sense the fine and subtle vibrations of Neptune, he brought submersion into and deepening of the spiritual problems and mysticism.

Now, under the influence of the successively intensity increasing vibrations of Pluto, not only the individual person, but the entire race, feels that something different is coming, something new, something yet entirely unknown; something which is not yet laid down in books; something of which one part of the people is afraid and which the other awaits with longing—the new age.

Every human being living will be convinced that he harbors in himself not only earthly things, but also heavenly gifts. Pluto breaks with the materialistic, mammonistic spirit of time and leads us across into an idealistic epoch with higher and purer aims for mankind. But this way leads through struggle and sacrifice because Pluto is struggle, a struggle until purification is reached.

Pluto is the borderline between two events—the decline of the old and the rise of the new era—the instigator of the turn in world events. But because Pluto is a planet which knows no pardon, no hindrance and no delay, which pursues his way calmly, its influence is consequently not felt in a soft manner, but on the contrary, it is very forceful and brutal. Pluto's destiny is to clean up the old and to march before the new era in a new form with its second luminous face of Janus proclaiming and bringing happiness and joy. . . .

If Uranus is the awakener, Neptune the perceiver, then Pluto is the ful-fillment of the new spiritual era. With the discovery of this trans-Neptunian planet, the new spiritual epoch entered upon its deciding stage. The first connection that Pluto formed after its discovery, with another great planet, was the square to Uranus.

This aspect I have classified as the aspect of the turning point of the world, and this has caused such upheavals throughout the whole Earth in the last few years that the world has been like a boiling kettle with events chasing and overtaking each other. Yet today we feel the aftereffects of this signifi-cant aspect. The human being is hardly able to catch his breath as he staggers from event to event, carried away and held in the grasp of the weight of the experiences. . . .

Pluto is giving the Earth a new face and calls forth deep cutting changes, especially in culture. He helps the human being in his struggle with the fet-tered energies and discloses new, heretofore unknown, forces of nature. The utilization of the heat of the Sun, of the slumbering force in the unsmashed atom, of the enormous reserves of earth-heat inside the earth, of the gigan-tic energies which express themselves in the movement of the waters of the sea (ebb and flow of the tides), to make use of the volcanic forces of the earth, the experiments with transmutation of force, and also all more or less gigantic daring and revolutionary projects of our time, can all be classified as the outpouring power of Pluto. . . .

7

The Pluto/Persephone Myth

Evoking the Archetypes

GARY ROSENTHAL

The myth of Pluto and Persephone begins on an ordinary day, in a flowery field. A field where Persephone is picking flowers with a retinue of other virginal goddesses. What could be more pastoral and innocent?

One of the flowers seems especially luminous and alluring to Persephone (as it had been placed there by Gaia, an even older Earth Mother than Persephone's own mother, Demeter). And in picking this flower "arranged" for her—a hundred-headed narcissus—a chasm opens beneath her from the flower she's just unearthed. And in the next instant, she *herself* is plucked—raped, and carried down into the underworld by an invisible god—He of Many Names: Pluto, the Greek god of death. . . .

Though there's no one "right" way to read any myth, I find the evocative metaphors at the myth's beginning to include: Persephone's naïve lack of awareness of the depths lying beneath her, and what these depths contain. Secondly, the shocking *loss* of that innocence (and the suggestion that there is a primordial intelligence in cahoots with Pluto—Gaia—that doesn't *want* her—and by extension, *us*—to remain so innocently naïve toward what is invisible and deep). And lastly, that Persephone enters her underworld marriage to Pluto as an *involuntary descent,* much as our own deepening may be initiated by traumatic events that involuntarily arise, grab us, and pull us down—or these days, bring us to therapy.

In retrospect, such traumatic events may later come to be viewed as "benevolent seductions"—into depth itself. This may explain why Pluto was sometimes depicted as having two pairs of arms—one pair that seemed threatening, and the other in a gesture that seems *welcoming* to us. But from Persephone's more naïve (*kore,* maiden, preinitiated) view, she isn't courted by her husband to be; she's ambushed, raped, *abducted.* And as the poet Robert Creeley once heard her say, *"O love, where are you leading me now?"*

Yet a virtue of such "involuntary descents" is that they can lead us to a depth we might not have reached in any other way. Suddenly, we discover *we're not in Kansas anymore,* not on the flat surface of the Great Plains—nor the flattened plane of life's mundane dailyness, where we often get to pick and choose, and without such deep, immediate, and unsuspected complications from what we pick.

Instead, something immanently shocking, overwhelming, is taking place, like getting a cancer diagnosis or being pulled down by a shark. We've been singled out, as if stalked by something that seems to have its *own* designs, its own desire for us. Something more powerful than ourselves seems to have us in its grasp. Perhaps we're finally close enough to an agency of Pluto that he can now confer his riches, the depth of perspective he would bring us to—albeit kicking and screaming.

I'm suggesting that whenever our traumas, terrors, and vulnerability involuntarily appear and seem too "overwhelming," we may have *become* Persephone, may be about to lose some portion of our naiveté, may be in the grips of some form of mythic initiation. And Plutonic initiation—Pluto's style of *eros*—is not only "overwhelming," or leading us to feel "carried away," but involves being *pricked, probed, penetrated by what has been hidden or feared.* (Maybe this archetype thus lies behind the contemporary, anecdotal reports of people being abducted—and probed—by aliens.)

The name Pluto comes from *Ploutos,* which means "riches." And we seem to inherit these riches not by anything that is given or added to us, but by what is revealed from *what he takes away* (whatever we have held most dearly, whatever we thought we couldn't bear to live without). And finally he takes *everything* away.

And in this way Pluto is not only the god of death, and the "deepest" god of the Greek pantheon, but the god of *deepest wisdom.* For as the god of death and impermanence, his deconstructive power is such that he can also destroy the cognitive and emotional obscurations that veil the deepest truth. (Of the god's many names, in India he was called *Shiva*). And without metaphorically dying to who we have naively been taking ourselves to be, there is no "second birth," no "resurrection" into a more essential form; the butterfly of the soul remains encased in some form of larval shell.[1]

Destabilizing visitations from below, that the ego may normally defend against, at other times *are* quite overwhelming. Consensual reality becomes a helpless spectator. We are dragged down into the underworldly spaces, the psychic holes of our deepest wounds and helplessness where we feel the most deficient, like a vulnerable maiden being ravished. There's often an edge of terror. Plutonic initiations up the ante, like enriched *Pluto*nium. They're powerful, powerfully deconstructive (*corrosive* to the ego; and so the ego feels them as *destructive*). We have been grabbed by something that could destroy what we've known as ourselves, and all our best laid plans . . .

For these *terrifying* reasons, in the ancient world Pluto was usually not spoken of directly but instead referred to by *euphemisms,* as if even speaking his name might attract the god's attention, might lead the god of death to come for us. One of the commonly used euphemisms for Pluto was, in fact, "He Who Comes For Us All." Yet another was, "The Good Counselor" (a nuance reflected not only by Greek mythology, but also in Toltec shaman-ism when Don Juan Matus tells Carlos Castaneda to always keep Death as a counselor over his left shoulder). If Oedipus wasn't as universal as Freud had thought, Death shows up everywhere, comes for us all.

But as the Greeks seemed to know, Pluto is also the deepest of counsel-ors—a very demanding one, though: a real hard ass who won't let us get away with anything but the deepest truth—*which is all that might remain,* once our deconstructive initiation has gone deep, once Pluto has fully had his way with us.

And when Pluto is "doing therapy" with us, it's like he first drags us into the neighborhood we were the most frightened to enter, and then if we're lucky, or have the wherewithal to make use of the teaching he's offering, we might come out of that neighborhood transformed, no longer so afraid of the thing that used to scare us. He is a "good counselor" in that he first shows us what is not an adequate enough refuge, what is not a large enough peg for us to hang our hat. The world, it turns out, is deeper than we thought. And as Persephone discovered, maybe we are too.

It's like our brush with Death, our death or near-death experiences, the encounter with our deepest fears—our "abduction into ultimate depth"—can show us what's really important in life, what's most deeply true, what can be relied upon, and what not. A great light might appear. Goethe's last words: *"mere licht, mere licht!"* ("more light, more light!"). On our deathbeds, a deeper wisdom may beckon—as our vanities flee, like rats leaving a sinking ship.

Psychotherapy itself may be Plutonic, and like gambling or prostitution, an "underworld activity." Or I find something depth-conferring, unpoliced, and enlivened in conceiving of it in this way. And in fact, the night before C. G. Jung hung out his shingle to practice psychotherapy, he had a dream that he was opening the largest *brothel* in the world.

In important ways, Pluto and Persephone inform the imaginal backdrop for *depth psychology*. If the more unsavory manifestation of Pluto is an "underworld figure" (like a mafia don, a predatory psychopath, or a lurking pedophile or rapist), then the more noble side of the archetype is the *psychopomp,* a receiver or guider of souls—those more welcoming arms. But the god of the underworld is more an impersonal receiver of souls, though, harvesting a new generation of human beings every twelve to thirty years, which astronomically is the time it takes Pluto's irregular orbits to complete—and in astrology this is sometimes thus thought to be what determines "a generation." This isn't a warm and fuzzy deity—whereas the compassion of his wife Persephone attends more to human side of death's bewilderment and grief.

Though both the most and least successful of therapies inevitably end, the soul's deepening seems a continual or *cyclic dying*—to whatever we have last taken ourselves to be; and thus *has no* conclusive ending. And the same might be said of the relationship between Pluto and Persephone, which endures forever—through Persephone's eternal and bi-polar swings, her yearly cycling up and down, her transits between earth and the underworld. The relationship between them endures—even given its rude beginning—that matter of the rape. (Though when gods have sex we may need to read between the lines, or note the specific nuance of their style of loving, each of them different in important ways.)

The Plutonic *psychopomp* as therapist, though, is but one half of a couple who meet behind closed doors, hidden from view, like in an underworldly vault—or an illicit tryst that is nobody else's business. What they say and do there is "confidential," privileged information. Like what happens in Vegas *stays* in Vegas.

But even if one member of the couple at times is the *psychopomp,* still it takes two to tango, and in any moment the music could change, and suddenly the other partner begins to lead, to take them deeper. These confidential visitations, these cyclic encounters with each other, may lead them each to a dissolution into the same depth, and by doing so, can confer the sovereignty of that depth. For what is pricked, probed, and uncovered may become an initiation into fearlessness, a fearlessness toward the depths of feeling, the depths of grief, and depths of the psyche that are even deeper than death.

In cultures *not* mythically dissociated, important myths have commonly been enacted as *initiatory rites*—so that their metaphoric themes *don't* become enacted (neurotically) as *symptoms*. And as a psychotherapist working in a mythically dissociated, narcissistic culture, I've attempted to offer at least some initial approximate of this.

Amongst the most profoundly transformational of such rites were those celebrated at Eleusis (the Eleusinian Mysteries)—rites which were enacted for nearly two thousand years, and which were based on the myth involving Persephone's abduction. Unfortunately, a more nuanced understanding of the rites that comprised the Eleusinian Mysteries have largely remained an object of *conjecture*—because revealing the details of these rites was *punishable by death*. And in this way, these Mysteries have largely *remained* a mystery.

For two thousand years though, we *have* retained a similar rite celebrating a mythic death and reappearance—the Catholic mass.[2] But a problem with ritual—a problem encountered in any religion, or with any spiritual practice, really—is that the ego can get a hold of almost *anything*, such that things can become routinized, wrongly-vantaged, distorted, and thus no longer as able to effect a profound shift in awareness.

In this light, it's hard to know what *kept* the mysteries of Eleusis so vital and psychically charged for such a long time. Undoubtedly, the injunction against revealing the details of these rites must certainly have *helped* to preserve their numinosity—but can't in itself fully account for the profound impact of these rites. Recent scholarship, however, has plausibly suggested that this may have been aided by the ingestion of psycho-active ergot, a fungoid growth found on grains—the study of which later led the Swiss chemist Albert Hoffman to first synthesize the hallucinogen LSD.[3]

What scholars have more commonly agreed upon is that the Eleusinian rites seemed to greatly relieve their celebrants' *fear of death,* as if providing a perspective—or the pointing out—of what might be embraced or accessed in moments (such as death) when all else is being taken away. In our own culture, the *loss* of such a rite, or the failure to create one, has left us with a fearful lack of orientation in facing death, a lack of orientation toward what dies and what doesn't—a lack of orientation toward what we most deeply are, and are not.

The encounter with death/Pluto can be frightening, and in different ways for different people—for in a mythically dissociated culture, a culture that has no rites that might better prepare us for death, death can be the ultimate boogie man. But Persephone's abduction by the god of death is also an initiation into mystery, one in which she is penetrated by—and taken to—a greater depth. (Like the old Chinese Zen masters such as Han Shan, who take on the name of the mountain where they live, Hades—as Pluto was also known—names both an invisible god and the realm where he dwells.)

Persephone's descent into Hades is at once the loss of her familiar vantage, her familiar world and the loss of her accustomed identity. For she must suffer the separation from her mother Demeter, and the loss of her virginal, maiden perspective (which had been naively unaware of the depths lying *within* and *beneath* her) in order for her to inherit her own sovereignty, and to reign as a co-monarch of depth—no longer as a vulnerable *victim* with no agency, but as the *Queen* of the underworld.

And as such, in her matured version, hers is the mythological perspective capable of compassionately welcoming others, who in facing death are also involuntarily entering the mysterious realm where she's come to reign. In our modern world, hospice workers may be the embodiment of *this* aspect of Persephone—just as anyone who has been raped, molested, abducted, or has experienced an involuntary, terrifying descent into a greater depth may have embodied Persephone in her earlier, *kore* form.

As a culture, we've been *lacking* something like this: mythic rites that can lead to a shift of vantage. And a transformation of whom—or what—we have taken ourselves to be. For when a culture lacks such rites and teachings, people are more likely to remain in the collectively-shared condition of "taking themselves to be what they are not." In this way, there is not only a lack of orientation in facing death, and a lack of awareness of what *doesn't* die, but (from a Plutonic perspective) a collective case of "mistaken identity."

Our culture itself now largely abides in this mistaken, un-initiated condition: a condition where the self-image is what people embellish, attempt to

market, and naively take themselves to be. In the process we become dissociated from the depths of *being* itself. With this loss of depth—this *absence* of
Plutonic vision—"the butterfly of the soul" remains encased by something
akin to a larval shell. And the best term for this uninitiated, en-shelled condition is *narcissism.*

And so, perhaps it's no accident that the myth of Narcissus *ends* with a
depth-vision, a vision that leads to his metaphoric *death,* and to an equally
metaphoric *flowering.* For in dying beside a lucidly-clear reflective pool into
which he had been gazing, Narcissus then morphs into a narcissus flower, a
perennial that returns each year as the harbinger of spring. While the myth
of Pluto and Persephone (which also reflects a cyclical process of death and
rebirth) *begins*—as did this writing—with Persephone picking a version of
this very flower.

Notes

1. For a fuller exploration of the linked myths of Pluto/Persephone and Nar-
   cissus/Echo, see my forthcoming *The Death of Narcissus: and Other Heresies for
   an Age of Narcissism.*

2. It was due to the rise of Christianity, in fact, that the Eleusinian mysteries
   were closed down in 392 C.E by the Christian Emperor Theodosius I, who
   saw these rites as inspiring resistance to Christianity. Within four years the
   temple of Demeter and every sacred site in Eleusis had been sacked, leaving
   behind only ruins and rubble, where for nearly two thousand years transformational rites had been celebrated.

3. See R. Gordon Wasson, Albert Hofmann, and Carl A. P. Ruck, *The Road to
   Eleusis: Unveiling the Secret of the Mysteries* (Berkeley: North Atlantic Books,
   30th anniversary edition, 2008).

8
....

Old Horizons

THOMAS FRICK

In the summer of 1959 my family moved from Arkansas to Michigan, from a dry, flat land surrounded by cotton fields, where a rare dusting of snow over dry grass stubble was ideal for a half-hour's body-stenciling of angels before it vanished like dew, to a climate of lingering dirty accumulations of ice and frozen crust where I was always cold. In the fall of that year I entered third grade. Soon afterward two friends and I inaugurated a space program alongside that of NASA, which had been established a year earlier. Intending to send a twelve-foot rocket to Venus, we sent a letter to the new space agency and received helpful photographs of existing rockets and missiles: *Atlas, Thor, Juno,* and *Titan*. Since our space program was also a secret society, I declared that, as an initiation, the members be able quickly to recite the names of the nine planets in reverse order: Pluto-Neptune-Uranus-Saturn-Jupiter-Mars-Earth-Venus-Mercury. Why did I put Pluto first, the farthest, coldest, smallest, most mysterious orb of the solar system, the one with the longest and strangest orbit? Maybe in a cold climate I was seeking a trajectory from the wintry dark toward solar sustenance. Something charmed and called to us from that black distance, "the nearest faraway place" (the title of a 1969 Beach Boys instrumental). The fact that Pluto moved closer to the Sun than Neptune for part of its orbit was important: obviously it was trying to say something to us. Twenty-nine years previously (a preposterously short span, it seems now, looking back), Pluto had been identified and named. By the time we started our space program the Russians had already launched the first satellite, put the first living creature in orbit, detected radiation from the Van Allen belts, reached escape velocity from the Earth and initiated translunar injection, orbited the Sun, crashed a ship into the moon, and sent back the first photos of the lunar far side. These to us highly admirable accomplishments radically undermined the Manichean Cold War ruses of the day.

One night that year or the next I had a dream of Pluto, more indelible than powerful, in that it was accompanied by no great emotion, but my memory of it today remains as clear as it was fifty years ago. I was naked and alone on a small frozen planet, a featureless light-gray ice field like a scuffed-up skating rink, extending as far as I could see in all directions. Sparse winter scrub and bare trees occasionally interrupted the mostly barren ice plain. The planetscape was moderately bright, like a lightly overcast winter noon, the sky almost indistinguishable from the surface. It was cold but not too cold; I wasn't shivering, there was no wind. I wasn't lonely. The main attraction of this personal Pluto was that I could urinate and defecate anywhere, just like the few shaggy wolf like creatures errant on their own business on the trackless ice all around me. Over the years I've been completely uninterested in (not resistant to) interpretations of this dream, which feels less like a text than like primal cosmological data.

Since Pluto's anomalous and highly elliptical orbit is inclined to the ecliptic by 17 degrees, it seems misleading to say that it "crosses" that of Neptune, but its trajectory takes it closer to the Sun for twenty years of its 248-year orbital period. From January 21, 1979, to February 11, 1999, it was the eighth planet out; had our space program survived we'd have altered our initiation rite. Some snapshots taken by Pluto during 1990, the year when it came closer to Earth than it will be for another full orbit: the first McDonald's restaurants are opened in Russia and China; the Communist Party of the Soviet Union votes to end its power monopoly; the largest art theft in U.S. history takes place at the Isabella Stewart Gardner Museum in Boston; a decade after completing its mission by exploring the Saturnian system, *Voyager 1* takes the famous "Pale Blue Dot" photo of Earth from 3.5 billion miles away; the Hubble Space Telescope is launched; the World Health Organization removes homosexuality from its master list of diseases; East and West Germany unite; Iraq invades and annexes Kuwait, leading to the First Gulf War; the Channel Tunnel between England and France opens; and the first web page is created.

Also in 1959, the Egyptologist and alchemist R. A. Schwaller de Lubicz, in conversation with his then student André Vandenbroeck, conveyed a traditionalist view of the solar system in the cosmic hierarchy: "We cannot deny the existence of horizons whenever distance reaches its limit. That goes for the horizon of reason as well as for the natural phenomenon . . . The solar system is man's horizon. That does not mean that there is nothing beyond. Simply that what is beyond is not of his nature, not on the scale of his vision. Teleologically, the sun is the end of man. It is folly to think beyond. It will lead to nothing, and it might destroy what is."

The Christian esotericist Rudolf Steiner elaborated this traditionalist vision in several lectures given from 1907 to 1922: "The ancients rightly considered Saturn the most distant planet . . . Our present Saturn received its name in ancient times when the wise ones could still give meaningful names to things. It was given its name out of its very nature. Today, this is no longer done. From the standpoint of materialistic astronomy it is quite justified to add Uranus and Neptune; but they are only members of the solar system in an astronomical sense. They have a different origin, they are foreign bodies that have become attracted and attached to our system. They are invited guests, and we are right to omit them. There were beings at the very beginning of our Earth who were scarcely fitted to take part in further development, who were still so young in their whole evolution that any further step would have destroyed them. They had to receive a sphere of action, so to speak, on which they could preserve their complete youthfulness. As Jupiter contracted, the beings who withdrew pressed together something that has nothing to do with our evolution; it was essentially related to the withdrawing beings. Thus, first Uranus was formed and later, during Mars evolution, Neptune arose. The names Uranus and Neptune were not appropriately chosen as had formerly been the case, though some meaning remains in the name Uranus. It was given at a time when an inkling of giving the right name still survived."

Steiner died in 1925 and offered no comments about soon-to-be-discovered Pluto. Uranus, as it happens, lies at the edge of naked-eye visibility, though it was not recognized as a planet in that fashion. It was first noted by William Herschel on March 13, 1781, during a telescopic sky search, and when four nights later he saw that the object had moved, he determined that it was a comet. The idea of adding a new planet to the five of unrecorded antiquity was not an obvious one, and even though its orbit was soon seen to be noncometary, it was six months before its planetary status was generally accepted. Among the names originally proposed: Georgum Sidus (the Georgian star, after King George III), Herschel's Planet, Hypercronius, Cybele, Astraea, and Minerva.

Unexpected movement in the orbit of Uranus gave rise, through calculation, to the discovery of Neptune. After a few decades of refined observations it became clear that neither Uranus nor Neptune moved in they way that they should, so another unknown object was postulated, eventually leading to the discovery of Pluto, though subsequently it was determined that Pluto was too small to have had the requisite effects on Neptune and Uranus.

In 1923 the journalist, playwright, and political operative Dietrich Eckhart recorded the following comment by Adolf Hitler, to whom he was a mentor: "Yes!" [Hitler] cried. "We've been on the wrong track! Consider how an astronomer would handle a similar situation. Suppose that he has been carefully observing the motion of a certain group of celestial bodies over a long period of time. Examining his records, he suddenly notices something amiss: 'Damn it!' he says. 'Something's wrong here. Normally, these bodies would have to be situated differently relative to one another; not this way. So there must be a hidden force somewhere which is responsible for the deviation.' And, using his observations, he performs lengthy calculations and accurately computes the location of a planet that no eye has yet seen, but which is there all the same, as he has just proved. But what does the historian do, on the other hand? He explains an anomaly of the same type solely in terms of the conspicuous statesmen of the time. It never occurs to him that there might have been a hidden force that caused a certain turn of events. But it was there, nevertheless; it has been there since the beginning of history. You know what that force is: the Jew."

In 2006, seven years after Pluto resumed its role as the most distant planet, the International Astronomical Union held its annual meeting in Prague and took up the brewing question of its status as a celestial object. The Kuiper Belt, extending outward from the orbit of Neptune, was now understood to be populated by many smaller icy bodies, thought to be remnants of planetary formation, and Pluto was more like these, it was proposed, than it was like the other planets. In Resolution 5A a "planet" was defined as a celestial body that (a) is in orbit around the Sun, (b) has sufficient mass for its self-gravity to overcome rigid body forces, so that it maintains hydrostatic equilibrium (a nearly round shape), and (c) has "cleared the neighborhood around its orbit." A "dwarf planet" shared characteristics (a) and (b), but its self-gravity has not cleared the neighborhood around its orbit, and (d) it is not a satellite. On this resolution the votes were not counted, but it passed with "a great majority." Resolution 6A, "Pluto is a dwarf planet by the above definition and is recognized as the prototype of a new category of trans-Neptunian objects," was passed with 237 votes in favor, 157 against, and seventeen abstentions, which is less than two-thirds of the voters approving of the official reclassification. Alan Stern, head of the *New Horizons* mission, has called the definition "absurd," and lamented that "less that 5 percent of the world's astronomers voted." Though it refers to gravitational effects, the phrase "cleared the neighborhood" sounds, in human terms, like some kind of bullying, segregation, urban renewal, or police-state tactics, and Owen Gingerich, astronomer emeritus at Harvard, is among many scientists displeased with the fuzziness of the language and the concept. He also regarded saying that a "dwarf planet" was not a "planet" as "a curious linguistic contradiction." A significant number of astronomers continue to think that Pluto and similar small objects should be classified as planets, because they have, in many cases, cores, geology, seasons, moons, atmospheres, clouds, and polar caps.

Pluto hasn't changed, of course; we have. That the construction of our classification systems reveals as much about our thinking as it does about the objects of our thought was colorfully underlined by Michel Foucault when, in his famous preface to *The Order of Things,* he cited a (fictional) Chinese encyclopedia alluded to by Borges, which divided animals into fourteen categories, such as "those that belong to the emperor," "embalmed ones," "those drawn with a very fine camel hair brush," "et cetera," and "those that, at a distance, resemble flies." What we "decide" Pluto "is" will have an effect on what we know about it. Voting, whether representative or not, will only take us so far, yet how else? That is, at root, the epistemological dilemma of our age.

We presume, by and large, that things have the right names. On March 14, 1930, Venetia Burney, an eleven-year-old Oxford schoolgirl, suggested the name Pluto for the newly discovered planet. "I was quite interested in Greek and Roman myths and legends at the time," she later said, also recalling how she and her schoolmates placed lumps of clay in the park, proportionately sized and distributed, to represent the solar system. That morning she was at breakfast with her grandfather Falconer Madan, who saw a newspaper article about the discovery and remarked that the planet had not yet been named. Burney suggested Pluto. Mr. Madan told an astronomer friend of his at the University of Oxford, who thought it an excellent choice and telegrammed the Lowell Observatory in Flagstaff, Arizona, where the discovery had been made. (Based on an episode in the *Iliad,* Venetia's great uncle Henry Madan had suggested the names Phobos and Deimos for the moons of Mars.) Burney heard nothing more on the matter for more than a month. On May 1, 1930, the name Pluto was unanimously adopted. (That the first two letters were the initials of Percival Lowell, the astronomer who had predicted the existence of a trans-Neptunian planet but had not lived to see its discovery, was thought to be a factor.) When the news went public, Mr. Madan rewarded his granddaughter with a five-pound note. "This was unheard of then. As a grandfather, he liked to have an excuse for generosity," Venetia recalled.

Percival Lowell's widow, thinking that the old gods were "worn out," wanted to call it Planet X: "I do not like Pluto for the name."

Pluto's oldest name was Hades, thought to come from *aeides,* the unseen, or from *eidenai,* signifying knowledge. Feared by gods and men who refuse to utter his name, he is also known by epithets: presider over meetings, fetcher of men, receiver of many, giver of good counsel. He is the son of Cronus and Rhea, brother of Zeus and Poseidon. Having deposed Cronus, the three brothers cast lots for the kingdoms of the heaven, the sea, and the infernal regions. The last went to Hades, where he rules with his abducted wife, Persephone, over the other powers below and over the dead. He is stern and pitiless, deaf to prayer or flattery, and sacrifice to him is of no avail. Only the music of Orpheus has power over him. This was revealed when Orpheus charmed Hades into returning his wife, Eurydice, from the underworld. Hades has a helmet, given him by the Cyclopes, that makes him invisible. Because of his generosity toward Eurydice and his love of Persephone, the idea of his character underwent a radical change around the fifth century BC, under the influence of the Eleusinian mysteries. Instead of the life-hating god of death, he was now seen as beneficent, the bestower of things that grow from the earth, metals, minerals, and other blessings produced below ground. In this aspect he was called Pluto, from *plouton,* giver of wealth, and at most of the centers of his cult he is so worshipped.

The initial 1942 report on the discovery of the element plutonium was kept secret until after World War II; it was finally published in the *Journal of the American Chemical Society* in 1948, which is where the names *plutonium* and *neptunium* were both revealed. In his autobiography, Glenn T. Seaborg tells more about the naming of plutonium: "It was so difficult to make, from such rare materials, that we thought it would be the heaviest element ever formed. So we considered names like extremium and ultimium. Fortunately, we were spared the inevitable embarrassment that one courts when proclaiming a discovery to be the ultimate in any field by deciding to follow the nomenclatural precedents of the two prior elements. A new planet had been discovered in 1781 and, like the rest of the planets, named for a Greek or Roman deity—Uranus. A scientist who discovered a heavy new element [number 92] eight years later named it after the planet: uranium. The planet Neptune was discovered in 1846, so Ed McMillan followed this precedent and named element 93 neptunium. Conveniently for us, the final planet, Pluto, had been discovered in 1930. We briefly considered the form plutium, but plutonium [94] seemed more euphonious. Each element has a one- or two-letter abbreviation. Following the standard rules, this symbol should be Pl, but we chose Pu instead. We thought our little joke might come under criticism, but it was hardly noticed."

Precovery: before Pluto was identified by Clyde Tombaugh, just turned twenty-four, at the Lowell Observatory, in Flagstaff, Arizona, astronomers at other observatories unknowingly had taken sixteen previous photographs of it. The oldest was made at the Yerkes Observatory on August 20, 1909.

9
....

The Inquisition of Pluto
A Planetary Meta-Drama in One Act

JOHN D. SHERSHIN

The Cast:

PRO Lawyer for the Prosecution, philosopher and scientist (intense and opinionated)

DEF Lawyer for the Defense, philosopher and scientist (indignant)

CHAIR Spokesperson for the Review Board (a group seated to the right)

The Planetary Delegation:

LORD PLUTO a.k.a. your Grace, an extra tall robed man wearing cloak and hood (wearing platform or elevator shoes beneath his robe)

LORD JUPITER accompanied by the lovely IO and EUROPA, moons of Jupiter

LORD SATURN accompanied by TITAN, a muscular young man, large moon of Saturn

URANUS a tall robed man wearing a pointed wizard's hat sitting zazen style in a corner of the room

NEPTUNE a robed woman wearing a broad-rimmed knitted hat sitting next to Uranus in a corner of the room carrying a harmonium or keyboard equivalent

----

The Extra-Planetary Guest:

SIRIUS, HIS EXCELLENCY, REPRESENTATIVE OF THE STELLAR SYSTEM SIRIUS (bare, or bald headed, wearing a fine knitted, colorfully embroidered cape)

CHORUS Greek/Hindu Chorus (male and female wearing masks: Krsna, Hanuman, Tragic-Comic Masks in Greco-Roman style, and/or in chalk face)

The Scene:

A Hearing Room; somewhere in the Milky Way Galaxy. It is constructed in the shape of a rhombic golden hexahedron or the equivalent. A perfectly round table is placed square in the center of the room surrounded by asymmetrical elliptical stools.

The Chorus sits at the left side of the room in a semicircle. A small Review Board composed of men and women in business attire sits at the right, also in a semicircle. The Board can be heard mumbling or grumbling in low voices throughout the proceedings.

There is a delegation of the major planets and one seemingly minor but powerful planetoid. And others, important members of the Hearing, are seated at the table.

Uranus and Neptune are sitting together on cushions in a corner. They can be heard conversing / whispering throughout the proceedings. Neptune occasionally quietly plays the harmonium.

A large picture window in the shape of a trapezoid and framed with wood molding can be seen in the back of the room. It shows a deep space view of the solar system with the Sun and orbiting planets surrounded by a spherical haze. This is referred to as the "Big Picture." There are small sigils or glyphs representing traditional symbols of the planets hanging on the walls as well as several antique carvings.

The Chair of the Review Board steps forward and introduces himself.

Chair: Thank you all for coming, and to your prompt response to this request for a Hearing. I would like to introduce . . . *(individually introduces the Cast . . . Prosecutor, Defense Lawyer, Planets, etc.)*

Pro: Ladies and Gentlemen: We are all assembled here today to consider the case of his Grace, lord Pluto. Our *mission* here is to uncover his purpose, his motives, his physical character, the Whole Story of his place in this stellar system of our Galaxy, known as the solar system.

Def: He is only a planet or maybe a planetoid. Actually, a dwarf planet.

Pro: And with a name like his I would suggest much more. *(A beat.)* Your Grace, we will now proceed to probe into your shadowy realm. A descent into your realm of darkness . . . Do you hereby swear to tell the whole truth and nothing but the truth? No half truths or quarter truths. We measure truth here v-e-r-y carefully. Scientific. *(chuckles)* So help you God.

Lord Pluto: God is Dead!

Pro: Oh . . . No, No, No . . . I can see we are not getting off to a good start here. What will the authorities think?

Lord Pluto: Of what Authority do you speak?

Pro: Is it your assertion then, your Grace, that the great God of our forefathers, who the monkey man on planet Earth has worshipped for hundreds of years, is gone? Out of existence. Hiding, cowering in some shadowy cave in your dark realm. Some kind of ideological asylum. That now you, master of the souls of the departed, have captured the soul of God. And all the other ancient gods and goddesses?

Lord Pluto: All gone.

Pro: Does that mean Brahma, Yahweh and Jesus, Lucifer, and the Hindu monkey god Lord Hanuman, even the self-created Prajapati, are all gone too? And all the peaceful deities, and all the wrathful deities as well? . . . And then, what of yonder Sol? The great god of Life and Light, the very Center of the solar system?

Points to the Big Picture.

Lord Pluto: Merely a blast furnace of hydrogen and helium. Nothing more!

Pro: So that now you are in total control. The last one standing? What hubris, I say. A massive power grab!

Uranus: Yes, I recall being overthrown. And here I am. Gone but not forgotten.

Lord Saturn: Ha, Ha, Ha . . .

Uranus: I can see it posted all over Pluto's realm: God(s) Wanted Dead or Alive. Catch 'em if you can.

Neptune: If the God(s) didn't exist someone would have to invent them.

Lord Pluto: Well spoken Neptune, great Master of illusion.

Lord Saturn: Look out! I can see his Grace is playing hardball now.

Chorus: *(chants)* Invention and Deception
　　　　. . . at your Discretion
　　　　The gods live; the gods die
　　　　Can't live with 'em.
　　　　Can't live without 'em.

Pro: And furthermore I question what historic era you really think you reside in. The Ancient or the Modern?

Lord Pluto: Actually, I preferred the ancient Greek World, or even the Renaissance. Especially the late Renaissance. Those great historic periods.

Pro: Yes, indeed, your Grace, you and the Jesuits, I believe. Well, you were whistling a different tune then. God-fearing defenders of "humanity."

The famous or infamous Spanish Inquisition of late Renaissance Europe, when all manner of witches, heretics, and unbelievers were hunted down, interrogated, and then dealt with *(pause)* severely. You even look the part. Had some fun, eh! Now the tables are turned. Time for you to face the probe and the scabbard. Now's the time for your Inquisition!

Lord Pluto: There was more respect in those days for position and rank!

Pro: A little more respect for you, for all the gods, even I dare say for that great God of Everything that you *now* say has been banished? Are you to be-be-*lieved!* You are out of your ... century! Past your prime.

Def: His views agree with many modernist Earthlings of the twentieth century.

Lord Pluto: I could quote for instance: Mench-kin H. L., who says ...

Pro: *(interrupts)* That's pronounced "Mencken H. L." I believe.

Lord Pluto: In his "Memorial Service" he states that all the ancient gods, the gods of civilized peoples, once "theoretically omnipotent and omniscient and immortal, and all are dead." And goes on to say that given the blood and sacrifices offered them, that it is, no doubt, a good thing!

Lord Jupiter: A true Humanist!

Pro: A true "Memorial Service"! Someone who puts the human monkey man first and everything else last! And what of your Grace? Is it your business to scurry away all these fallen icons?

Lord Saturn: New and worthier deities demanding a greater and larger sacrifice of blood and treasure have taken their place. But that is not his Grace's responsibility.

Pro: Yes, indeed, was it not at the time of the very discovery of this far-flung planetoid, in the early twentieth century that the monkey man unleashed greater death and destruction than ever occurred on the planet. You might say the Earthling devolved to his worst chimpanzee behavior! A Final Solution alright! What does your Grace say to these charges?

Lord Pluto: I was always there. For billions of years. The monkey man did not yet acknowledge my presence.

Lord Saturn: Is that so?

Pro: And when he did acknowledge you, all Hell, excuse me, all Hades broke loose. If the gods be dead as you claim, then, by Symmetry, I declare the god of the dead is dead.

Def: Objection. Symmetry is not an admissible argument. It cannot be used here. Someone broke into Symmetry. Symmetry is broke! He's down on his luck. Damaged. Not reliable and even unstable. A nice idea, but . . .

Pro: Then partial or bi-lateral Symmetry? How about Dynamic Symmetry?

Def: No quibbling. It's off the table.

Pro: That so, well then perhaps death lies dying. Is that better, or worse?

Lord Saturn: Hey, what a concept! The very inverse of mortality!

Def: We also agreed that there would be no talk of mirror images, inversions, or any form of symmetry, and especially . . . any reference to immortality.

Lord Saturn: But how can it be avoided?

Chorus: It is Agreed, It is Agreed,

It is Decreed, It is De-Creed

. . . Indeed.

Pro: His Excellency the Representative from Sirius would like to speak.

Sirius: Yes, I say this has all gone too far!

A loud uproar among the delegates.

Def: Please Gentlemen . . . You will have your turn to speak, your Excellency.

Pro: Again, I propose that you are out of touch. Out of your . . . century!

Lord Pluto: Oh?

Pro: Yes, it appears the monkey man has opened a new paradigm, entered a new historical era. A different Worldview for the twenty-first century. It is called "post modern."

Lord Pluto: Did you say "post mor-dem"?

Pro: P-o-s-t m-o-d-e-r-n. Anything goes, sort of. No more dull drab modernism. No more nihilism. No more iconoclasm. The smashings over. Just like you, your Grace, boring Monolithic Modernism is finished. Tired of plain or plane. Let's "take a fancy." Get dimensional . . . Too logical, too rational.

Lord Saturn: The logical positivists. The modern philosophical movement.

Pro: Exactly, way over the top.

Lord Jupiter: *(Stands up, spreads out his arms and breaks into song.)*

Ya gotta have h'art, miles and miles and miles of h'art.

It's nice to be a genius of course,

But don't put that horse before the cart

First, ya gotta have h'art.

Lord Saturn: Where did that come from?

Lord Jupiter: *Damn Yankees.* A 1950s Broadway musical.

Uranus: *(to Neptune)* I remember that one with the sexy dame that played the lead.

Neptune: *(looks away disdainfully)* Clearly before my time.

Uranus: No way!

Pro: Please . . . I appreciate the enthusiasm, but no more spontaneous outbursts! As I was saying . . . Subjective and Objective, all rolled up together. Paleolithic, ancient Roman, medieval Europe, Hindu and Buddhist, all in the same basket. But no Aquarian or contemporary mish-mash, mind you. It needs to hang together. Do something. Say something. Tell a story. Preach a riveting truth. The gods live and the gods die. It's an everything stew. Just keep stirring the pot.

Chorus: A witches brew and stew

        A witches stew and brew

        A goblin's goo

        A magician's pot and cauldron

        . . . A magician's spot of . . . Whee-ew!

(The Chorus twist and turn while obviously holding their noses.)

Lord Saturn: How enlightened!

Lord Pluto: Bunk! . . . Now all the great philosophers are "hanging together", and "telling stories." The monkey man is forever revising his paradigms! And anyway, what's that got to do with me!

Lord Saturn: Just the facts, please.

Lord Jupiter: Now I can go for that. Your facts, or my facts!

Pro: I was simply trying to ex . . .

Def: *(interrupts)* Objection . . . Gentlemen, this is all very interesting, or not very interesting at all, *(angrily)* BUT WE JUST DON'T HAVE TIME FOR IT!! There is an important investigation before us, and we must proceed.

Pro: To the point, your Grace, is it true that this current probe sent by the monkey man for your examination is powered by one of the most dangerous and deadly radioactive elements created by the Earthlings, namely plutonium-238? The light and power of yonder Sol (points to the Big Picture) can barely reach your dark domain, so something else, more sinister has been devised.

Lord Pluto: Yes indeed, but of no consequence. And the risk is quite small. Why it's hardly a nuclear bomb or even some huge reactor.
Laughs convulsively.

Pro: Indeed, the risk is always small until something happens!

Lord Saturn: I barely escaped a few years ago during the Cassini probe.

Lord Jupiter: And me too.

Uranus: I have been spoken and accounted for. I've been named. I'm in my element. Uranium.

Neptune: And so am I. Neptunium. *(She rolls her head and eyes.)*

Pro: And does your Grace not have some kinship with this very plutonium?

Lord Pluto: They just borrowed my name.

Chorus: What's in a Name
      Been said before
      . . . said before
      The International Astronomical Union Knows
      The Naming Committee of the Astronomical Union Surely Knows. . . .

Lord Jupiter: There's more to a name, I would say.

Lord Saturn: Less, I would say.

Pro: And who your Grace, who on Earth, christened—forgive me—gave you your Epithet?

Lord Pluto: My Epitaph? The Earthling Disney, I suppose. No . . . it was a young lady admirer.

Pro: Is that so?

Lord Pluto: I have my charms.

Pro: Charms and Talismans, no doubt, to fool the foolish!

Chorus: No . . . No, No. It was the Fates.
      The Fates are in the Name
      The Names are in the Fates. Indeed.
      It is Decreed, Agreed.
      It is De-Creed. In Deed.

Pro: And in addition, there is the question of your orbit, your Grace?

Lord Pluto: And what of my orbit?

Pro: It's a bit irregular. More eccentric than that of Uranus! It resides in an entirely different plane from those other planets surrounding yonder Sol. Furthermore, you have flagrantly disturbed your neighbors Neptune and Uranus.

Chorus: *(chants)* More than Elliptical.

Elliptical and Eccentric.

Not one of us. Never was. Never was.

Eccentric. Not normal. Never was.

Banished . . . Banished

To the End, to the very Edge

Out to the very Edge

The Edge of the End!

Def: No more than the others here, he obeys gravity's Universal law of inverse square attraction while scribing his orbit 'round yonder Sol.

Lord Pluto: *(interrupts)* In fact, should there have been the rule of inverse cube, as one might expect in some multi-dimensional and non-Euclidean universe, I would have long since jumped out into the Beyond in some logarithmic spiral.

Pro: Instead of wriggling about in some eccentric orbit around yonder Sol.

Chorus: An Alien Intruder . . . an Alien Intruder

Captured from the Beyond . . . the Beyond . . . the Beyond

The Unknown . . . the Unknown

A Prisoner in Sol's World.

Def: Free Pluto, free Pluto from his Solar shackles, from his grim and ancient duties; from his co-dependence on the mad monkey man. Into the Dark Void and Unknown! Let him go.

Chorus: Free Pluto, Free Pluto,

Let him go, Let him go!

Free Persephone, Free the Captive!

Let her go, Let her go!

Lord Jupiter: *(to Lord Saturn)* I kind'a like that idea. Year-round harvest. Maybe no winter, too!

Lord Saturn: We should have ejected his Grace long ago when we had the chance!

Pro: Indeed, send him into deep space—where—he—belongs! . . . Evidence recently acquired shows that once a mighty and dreaded god, once thought a great planet even, you are really nothing more than a swarm of cosmic debris and paltry balls of ice! It was thought that your light was just the small reflection of a dark and larger globe. But no, you no longer even have the status of a planet. Dethroned. A mere planetoid. A dwarf. You're finished. Your reign of terror is over! And there is more to come . . .

(Lord Pluto pulls his hood tight over his head, and blinks spasmodically.)

. . . Indeed, Have we all not been hood-winked by your Grace and your ruthless grab for power! *(LORD PLUTO kicks off his platform shoes unnoticed and shrinks into a small crouched-over position.)*

Def: Objection. Lord Pluto remains the tenth most massive planetary system in Solar orbit. *(Points to the Big Picture.)* And has satellites, followers, I might add. Nyx and Styx and Charon and Kerebus.

Pro: An unlikely bunch! Some 3-headed jackal, the Night-time and a river of forgetting . . . of remembering . . . ?

Chorus: Nyx and Styx
Styx and stones
Dogs and drones . . .
Can break my bones . . .
Moans and groans . . .
But Names will never hurt me.
. . . Names will never hurt me.

Lord Pluto: You speak of Lethe and Mnemosyne, other unseen and yet to be discovered rivers in my realm.

Pro: And what of Persephone? Where is she? *(He looks sternly and accusingly at LORD PLUTO.)*

Lord Pluto: That's my business.

Pro: And in addition to this obvious "f-o-l-l-o-w-i-n-g" as the Defense would call it, His Grace has covert operatives, too.

Lord Pluto: Really!

Pro: You need not divulge. We already know. There are many others, be it hundreds just like you. What Earthlings call "plutoids." Puny paltry plutoids. All lodged in a huge circular ring known as the Kuiper Belt. You see, for centuries the great Lord Saturn paraded with cloak and hood with scythe in hand to scare the living with his spectre! The very the last word on Rings. Until recently in the course of history he ruled all boundaries, the very edge of the known solar system. But no more. Others, too, have Rings.

Uranus and Neptune are heard chuckling.

Neptune: *(spreading out her be-jeweled hands in front of her and speaking loudly and clearly)* And each Ring with a Name: "Courage," "Liberty," "Equality," and "Fraternity."

Lord Pluto: Say What?

Lord Saturn: You heard!

Pro: But now there is even a larger Ring. A Ring of Closure. The Ring-Pass-Not. The Kuiper Belt, I speak of, and its band of Plutoids.

Chorus: *(mockingly)* Hail Pluto. Bring his Scepter,
>Bring his Crown,
>Bring his Ring! . . .
>Hail Pluto, King of the Kuiper Belt!

Lord Pluto: *(Pulls off his hood and cloak, wearing a dark tight fitting garb with a black skull cap, pointed goatee and long curled and pointed shoes-Renaissance or Elizabethan style dress.)* Are you afraid of the Dark, of the Depths? Of things that glow in the Dark, of things that go Bump in the Night. Of falling into a Deep Dark Hole and waking in another World? Of buried treasure. But instead cling sheepishly to light and clarity, to the Light and Bright of yonder Sol. Always ascending, never descending. Always producing; Always reproducing. Go Underneath, Go Undercover. Where the dead are buried. Where ashes fall. A place for forgetting, and for remembering. Where seeds germinate and where organic meets inorganic. The fertile humus. Of finding Fossil's Fuels, or is it Fool's Fuel. (Leave it where it lays.) Be still and lie fallow. Come Underground to Pluto's Realm, or come to The Dark Edge at the End . . . Either way . . . Come, sit in the Shade and enter the Shadow World. Do not be afraid. Let a little Darkness into your life. Empty out into the Deep Dark of Space. No, no, bring it to the Light of Day! To Sol's World. *(Looks scornfully at the Big Picture.)* Or keep it in Mine! There is more to my realm than meets the eye . . .

Pro: And there is not much there to meet the eye!

(There is a long pause in the dialogue. Neptune quietly plays on the harmonium.)

Def: Yes, there remains a place for his Grace in Sol's world. He is *different*. He lives at the End, on the Edge, but he is *one of us!*

Pro: Maybe.

Chorus: The Dweller on the Edge
>The Dweller on the Threshold

lord Hanuman and the monkey man
Will soon find his secrets
Praise be lord Hanuman ...

Pro: And what do you see, your Excellency, Representative of the stellar system Sirius? Speak now!

Sirius: Thank you for inviting me to this Hearing today. As you are aware, we of the Stellar System Sirius are approximately eight light-years distance from yonder Sol's System. We have examined many exo-planets, many living planets throughout the galaxy by our advanced methods of science and technology, our long research into the great laws of this universe. We have evolved differently from those creatures on planet Earth, not unlike the evolution on those island continents on their planet, but more so ... Fortunately we have not been encumbered with the more disturbing, shall I say blood-thirsty characteristics of the monkey man on planet Earth. The Creator has endowed us with the ability to advance far into the worlds of mathematics, science, and our specialty: geometry, that wondrous description of the physical universe that we all share. Yes, we may not have some of the more endearing traits of the Earthling, so well demonstrated here by the charm and jovial personality of lord Jupiter. Sentiment and sentimentality are not our strong points. Maybe that is unfortunate for our species? Nor do we share the ponderous melancholy of lord Saturn or the cold and icy ruthlessness of his Grace. However ... We have carefully observed this verdant planet Earth, one of the few and finest living planets in the Local Region. For the past quarter of a century, our planetary system has consistently sent messages to the fields of Albion in the islands of Britain on yonder Earth, where once primitive Earthlings constructed great geometric structures. Sending vibrations of light, electromagnetic waves through the void of Minkowski space, nearly a decade in advance, which then interact with local atmospheric conditions on that planet to create geometric patterns of splendor and precision. Grand and profound symmetries! Some of construction unknown to the ancient Greek geometer Euclid, or the contemporary Mandelbrot.

Pro: Prove it!

Sirius: Is that really necessary. Is it not self evident and well documented!

Def: Thought it was the handiwork of some Oxford Don playing with some new top-secret gadgets.

Pro: "GPS" devices on silent motorized trampers. Creepy, in the middle of the night, too, with no headlights!

Sirius: There are also many coded messages, but I submit that the monkey man just DOESN'T GET IT!

Pro: Get what?

Sirius: *(looking frustrated)* Concerning the matter here before us: It has not been our position to banish lord Pluto and his fellows from our stellar system. We know naught of his arising, or comings or goings. No doubt we have been fortunate not to witness his dark denials and deadly dealings, nor his open quest of the Dark and Unknown Void. *(Points to the Big Picture.)* Furthermore, we see beyond planar 2-dimensional Rings and Orbits a large Spherical Cloud enveloping mighty Sol and his planets. A large enclosed ball, suspended in the Void. What Earthlings call the Oort Cloud.

Chorus: The Oort Cloud
        The Cloud of Oort
        The Cloud of Un-Knowing
        The holy 'rit of the medieval sage
        The Cloud of Knowing
        The Ghostly Cloud
        The Cloud at the End of the Edge,
        The Great Spherical Cloud, a Giant Crystal Orb
        The Cloud of Closure ...
        ... The Cloud of Closure
        The Monkey Man and Lord Hanuman
        Such clever Fellows
        Will discover many things
        Praise be lord Hanuman ...
        Praise be ...

Chair: *(Stands and addresses the gathering.)* Thank you, Thank you, Gentlemen and Women; Delegations from the solar system, and the Sirius Stellar System. Thank you very much for your attendance here. That is all we have time for today. The Board will ponder the testimony and evidence presented. Until further notice, the Hearing is adjourned.

Lights dim; curtain falls; loud mumblings and grumblings from the Review Board can be heard, fading away.

10

Pluto and the Restoration of Soul

STEPHAN DAVID HEWITT

Every exploration in the outer physical world is symbolic of our desire, often unconscious, to explore our inner world, the world of psyche. When Pluto was first discovered in 1930, humankind was already beginning to tinker with the atoms of uranium. That tinkering would, years later, result in the release of huge amounts of plutonium into our environment, an element occurring naturally in the Earth's crust only in very small amounts: 3.5 tons of it have since been released into the atmosphere by atomic testing. A very small amount of it, one-tenth of a milligram, is said to cause cancer. Pluto's discovery heralded the emergence of the great archetypal power that it symbolically holds into the hands of humans, as well as its great exploitation.

As usual for a race so young as ours, we poke, prod, dissect, explode, and misuse the elements of our environment until we discover (often too late) that we must be careful with them, or we could destroy ourselves, and all life on this planet. We are simple primates who are intensely curious and excited by our newly discovered toys, but lacking in foresight or objectivity. So now we are sending a probe into the deep reaches of our solar system to find out just what is going on with Pluto, which seems to be no longer a planet but a cluster of bodies, like atomic particles in orbit around each other.

Astrologically, Pluto is still a planetary body, and quite an important one, as its 248-year cycle around the Sun indicates major changes in different areas of our collective psyche. The astrological signs that Pluto transits through are relevant to those changes. In 2015, we are now in Pluto's transit through Capricorn, the sign it was in when the United States was just forming the Declaration that would create its separation from England in 1776. Capricorn rules the structures we create to organize our lives, most notably governments. Pluto will return to that same position in 2022, but in 2015 it will form a 180-degree opposition to the Sun in the chart of the United

States, and square (form a 90-degree angle) the United State's Saturn. When Pluto makes these kinds of aspects, it is a test of our country's maturity and a questioning of its national identity. Also, wherever Pluto transits, it rules transformation of our relationship to power. By the time *New Horizons* reaches its goal, the United States will probably be undergoing massive changes, which will affect the rest of the world as well.

Pluto's realm is darkness. It is the pit of the bowl of the crucible where the deep underground fire of transformation takes place. Pluto's territory is anathema to the sunny warmth of the ego. But it is power. Ultimate power. The kind of power that easily slips into megalomania if the ego gets hold of it and begins to identify with it as *my* power. One of Pluto's lessons in power is that it belongs to no one, that the power we hold is shared among all beings, all the creative forces of the known and unknown universes. The history of humanity can be seen as our trials in misunderstanding this universal truth. Yet the secret of Pluto's power is the knowledge of how shadow and light, when combined, produce the unification of the opposites, the balance point of yin and yang, the alchemical transcendence of the limitations of our physical form and reality. In short, Magic. Real Magic, not the flummery of Harry Potter, but the kind of magic that allows us to glimpse our reality from the perspective of the gods.

The truth of our world is that it is an extension of our psyche: every outer exploration brings us more into contact with the components of our deepest Self, the Soul. As above, so below; this is the truth of the psyche as well as the backbone of the art and science of astrology. As we send a mechanical extension of our eyes and ears to the farthest reaches of the known solar system, we also send our psyche there, to witness physical objects that will inevitably have their corresponding effect on the inner reaches of our consciousness. Like our tinkering with the atoms of uranium, we have little knowledge of where this exploration will lead us, or how it will cause us to evolve.

Pluto's realm, the underworld, is often characterized by and confused with the realm of hell, a fiery tortuous territory of eternal punishment, or a place of eternal sorrow. It is indeed the land of the dead. But archetypally, this land is a numinous land: it is the land of Soul. It is only dark to our mortal eyes, which cannot directly see into the immortal inner planes of the territory of the Soul. Thus Pluto is the bringer of light—the enlightening energy of Soul we must draw upon when the mundane worlds of ego and personality fail us, as eventually they must.

Perhaps the *New Horizons* probe will find a scientific anomaly of such magnitude that Pluto will forever point us into deeper, more infinite directions of inexplicable wonder. More likely, our clearer vision of Pluto will be a simply marker point in human history. As we face greater threat from the poisoning of our environment by humankind's deliberate ignorance fostered by greed, and the erosion of our stable climate and weather patterns, we may be forced into great changes in the way we do business here on planet Earth. Faced with cataclysm, we may have to drop our pettiness, our nationalism, our war budgets, and spend our money on helping out great numbers of people in need. This is how Pluto works when we fail to consciously recognize the need for change. Its effects are inexorable.

A very important astrological cycle will also conclude in 2015, the square (90-degree angle) between Uranus and Pluto, which began in 2012. Uranus in Aries, the sudden awakener, and Pluto the transformer last met in this hard aspect in the years 1932–34 (just two years after Pluto's sighting) and set the stage for large-scale shifts of power in our world. We are now at that launching pad again, with a shift in emphasis from Pluto in Cancer in the 1930s to now Pluto in Capricorn, where it was in 1776. (Due to Pluto's eccentric 248-year orbit, it passes through some signs much faster than others.) There is great speculation as to the future of the United States after 2022, when Pluto returns to its place in the national birth chart (right around Independence Day, 2022). A Pluto return heralds powerful, enormous changes. Doubtless this country will not be the same thereafter. Change requires destabilization and breakdown before the United States will let go of its fascination with itself as a world power. We can already see the seeds of this as we face our multiple trillion dollar debt, crying poor and cutting social programs while spending exorbitant (and unaccountable) amounts on war machines. This is the end, and the beginning of something new.

Pluto's appearance before our eyes will awaken deep stirrings in our collective unconscious, just as landing on the Moon in 1969 created a tidal wave of psychic energy as we looked back from the Moon and saw our tiny Earth floating in space. The Moon astrologically represents our emotions and feelings, while Pluto's effect is to deepen, magnify, and stir within us our most primitive and instinctual drives, the kind that erupt spontaneously as crowds, mobs, protests, or "flash mobs" and sparks of profound revelation. The Moon is a very visible force, Pluto's force is hidden deep underground.

Since the probe is speeding by at 36,000 miles per hour, it will only spend twenty-four hours taking pictures and measurements of Pluto, but even these glimpses will unlock wonder and imagination within our collective human mind. Symbolically we are trying to peer through the veil of the limits of our consciousness. It seems we are asking for a clearer vision of the dark god who rules over the powerful instinctual energies that are deep within us all.

Pluto's realm may be deemed dark and dangerous only because we have become so enamored of our egos that we refuse to see the limits of our power, to acknowledge our smallness in the face of the great unknown. If we are to learn from our indigenous ancestors and their practices, we can strengthen our container, our bodies, our communities, by calling upon the helping spirits of that realm, who were known to be readily available to us in times previous to our present Age of Reason. When we partner with the unknown, rather than fear it, we take a step toward the unification of the opposites, the small "I" and the larger "we," the individual and the collective, the conscious and the unconscious. When we release the hypnotic hold of the material world, the boundaries begin to blur, and we enter into the realm of the hidden forces that power the material world, call it spirit, or archetypes, or quantum energies. Our worldview shifts and we begin to see shapes in the stories of our lives that bespeak the pattern of Magic that flows through our days, through synchronicities, powerful dreams, by meeting the right people at the perfect timing in our lives.

The deep volcanic power within the earth need not destroy our lives if we listen to the earthquakes that precede eruptions. Similarly, if we pay attention to the signs and signals that our reality continuously sends us (when we are quiet and still enough to listen) we can create safety and refuge and see our way into the future. Pluto's darkness may be the doorway to discovering our eternal Self, to seeing beyond the ego's desires and prejudices, to the discovery of the guidance of Soul with all its thirst for experience, understanding, and knowledge of the great Universe we are enwombed within.

Our visit to that cold icy world may be the signal to humanity that it is time for us all to learn how to cozy up together and create mutual warmth in the sharing of our stories. That it is time to discover we all need all of us, together, in order to survive the vestiges of past horrors we have inflicted upon each other and our planet. Pluto strips away boundaries, nationalities, racial prejudice, everything we use to separate ourselves from others out of fear. It is the power of Oneness. It shows us how common the experiences of life are that we share, despite differences of language, color, or custom.

The essence of any long-term relationship is our ability to repair and continue. Out of the ashes of our self-destruction can come the restoration of our world, a rebirth of Soul. The synchronicity of the name of the *New Horizons* probe, as well as its timing in the history of the country that sent it, are harbingers of a new era of the humble awareness of our dark powers, the history of our self-destructive designs, and the beginning of the restoration of Soul upon this fragile planet we call home.

11
.....

Our Lady of Pluto,
the Planet of Purification

Jim Tibbetts

The myth called Our Lady of Pluto is a story in which a mysterious lady in a globe of light glided by the planet Pluto six times in 1917—something it remembers quite well. The planet Pluto has a consciousness as all planets have a consciousness, one that is holistic and integrated with a memory of past events and future possibilities. Pluto is a very important planet, if not the most important one, because it is the omega point of the solar system of Earth, and the reflecting point of purification.

Pierre Teilhard de Chardin, S. J. was a French paleontologist who lived from 1881 to 1955. He articulated a new dimension to evolution: a spiritual dimension. He taught that there is an unfolding of the universe as a physical evolution and alongside this is the unfolding of the universe as a spiritual evolution. The Jewish-Christian development over the last 3,700 years has thought in terms of a single linear development of their faiths. Teilhard started to see the universe in an integral way, not just in the fundamental subjective/objective dualism of most Western thought. Along with this is his central idea of a complexification of consciousness. The evolution of the universe is a deepening of human consciousness or intelligence that includes a deepening of interiority and spirituality. It is a biological-spiritual process growing together.

Human evolution continues as is but Teilhard had a sense that a deep change, like a transformation or conversion, at the level of being—a change of heart, a change of mind, a change of body—can take place in a human who learns to experience the universe with divine action. This becomes a mystical event in the life of the individual; it can be called mysticism or divinization. This also exists for a planet that evolves, has a planetary consciousness, a heart and thoughts.

Teilhard is known for his "omega point," which is the point the universe has become one with God. It is an event that the universe is moving toward in the future and that is exerting a force on the present. The Western mind usually thinks of the omega point in linear terms, but for Teilhard it is in holistic terms: the omega point is radiating back from the future into the present, thus the future is here and now. Planet Pluto is the "omega point" for the solar system; it is a reflecting point of Earth's consciousness and events, reflecting these events back to the Earth.

A person's positive and negative beliefs not only impact a person's health but also every aspect of his or her life. Teachers like Buddha and Jesus have been telling us the same story for millennia. Positive thoughts are a biological mandate for a happy, healthy life. In the words of Mahatma Gandhi:

> Your beliefs become your thoughts
> Your thoughts become your words
> Your words become your actions
> Your actions become your habits
> Your habits become your values
> Your values become your destiny.

The planet Pluto has thoughts and a consciousness, one that is holistic and integrated with beliefs and thoughts that are part of its history and destiny. Though it is cold and icy, dark and distant, it lives on with its planetary consciousness. One moment in history brings Pluto great joy: it is the presence of the bright luminous globe with a beautiful lady inside that floated by once a month for six consecutive months.

Pluto is the planet of purification. It is a mystery how it became a planet of purification; many reasons have been given but all fall short in the ocean of speculation. Perhaps since it's the farthest planet from the Sun, the main instrument of purification, its hermetical lifestyle throughout the ages sublimed itself as the omega point for absolution, a cleansing of consciousness.

Pluto, a small planet about one-sixth the size of the moon, is made up of mostly rock and ice. Discovered in 1930, Pluto was originally classified as the ninth planet from the Sun. After 1930 many eyes started looking at the planet Pluto and its consciousness became aware of those humans looking at it from the planet Earth. Yet, Pluto will always remember the globe, with the beautiful lady of light, which passed by it six times in 1917, an event that elevated its planetary consciousness forever.

Putting the speculation aside let us turn toward the science and faith of purification in one of the Earth's most ardent promoters, the Judeo-Christian tradition. The spirit of purification first started with Adam and Eve. Tens of thousands of years went by and the caveman never achieved this art form, this wisdom of deep space. Abraham, Moses, and the Jewish tradition picked up on this science and art form and attempted to live it out. At the end of the lineage of the prophets came the prophet Rabbi Jesus; his first words in his mission were, "Purify thyself, for the kingdom of God is at hand." His mother radiated purification itself. She lived on throughout time and came back to Earth to spread a purified lifestyle.

Planet Pluto does not think like humans do, but experiences emotions and images and power and beauty in a two dimensional way. All things that are being purified come to Pluto and bounce off this omega point, reflecting it back to its source as a purified experience or image.

One six-month period on the thirteenth day of every month from May to October it experienced a luminous globe with a beautiful lady floating by it in its orbit around the sun. The light from it radiated a purity and compassion it had never experienced before. The purity that radiated out from the event, taking place on Earth in Fatima, Portugal, was bounced back to the pure core of the planet Pluto. This event was called an apparition of this beautiful Lady; the story from Pluto's memory follows thus:

First Apparition: the orbit of Pluto first sees this glowing globe pass by

When the prayers are finished, the three children were just about to resume their games, when suddenly a bright flash of lightning blinded them. Afraid of a gathering storm in spite of the cloudless sky, Lucia (the visionary who spoke to the Lady) made up her mind to go back home. But scarcely had they taken a few steps in the direction of a young oak tree when a second flash of lightening, brighter than the first, brought them to a standstill, frightened and trembling. On their right, in the blinding light, a young lady (so Lucia describes her) of dazzling beauty appeared before their timid eyes. With a motion of her hand, she reassured them: "Do not be afraid, I shall not hurt you." And so they remained there in ecstasy contemplating it. The beautiful lady spoke to Lucia for a minute and then rose up into the air floating away on that day, May 13, 1917. This radiated out to Pluto who beamed it back.

Second Apparition: Pluto experiences purity from the globe as never before

The globe of light descended again from the heavens and Pluto rejoiced. This first apparition was spoken about to everyone's amazement and only about fifty people betook themselves to Cova da Iria, most of them through curiosity for the second apparition on June 13. Suddenly Lucia made a gesture of surprise: "Look," she said, "that was a flash of lightning, the Lady is coming!" Throughout the apparition, the witnesses, who did not see the Vision nor hear what she said, noticed that the sunlight had lost some of its brilliance, and that the atmosphere had become yellow-gold. The others heard something that sounded like a very gentle voice but did not understand what was said. "It is like the gentle humming of a bee," someone whispered.

When the apparition ended, the curious onlookers approached the oak to examine it. It was covered with long new leaves, for it was June. They found that all the leaves of the little tree were bent over toward the east, as if the hem of the Lady's mantle had brushed over them as she turned and moved away in that direction. They had heard Lucia's questions but had neither seen nor heard the Lady. Pluto listened to words but did not understand.

Third Apparition: the luminous globe with the lady passed by Pluto with colors superb

The fifty people who had witnessed the apparition of June 13 spread the news of it throughout the surrounding districts, and on Friday, July 13, more than five thousand were hurrying on their way to Cova da Iria. The sun turned hazy and a refreshing breeze began to blow. The silence of the crowd was impressive. Then they began to hear a hum as of a gadfly within an empty jug, but did not hear a word.

Looking closely, they saw something like a small grayish cloud hovering over the holm oak. During the apparition the spectators noticed that in addition to the sunlight's becoming dull, a little white cloud surrounded the children and covered the scene of the apparition. Some of the bystanders noticed that the light of the sun seemed dimmer during the following minutes, through the sky was cloudless. These images impressed themselves on Pluto.

Fourth Apparition: the globe passed by Pluto filling its heart-center with love

On Monday, August 13, the newspapers estimated the crowd at Fatima at eighteen thousand persons. At the exact hour of the apparitions, the attention of all was drawn to the tree by a loud explosion. Then followed the flash of lightning, which usually announced the coming of the Lady. The three child visionaries did not show up since they were taken away and interrogated by the police.

From the midst of the clear sky a clap of thunder broke, and a brilliant flash of lightning shot through the heavens. The sun began to lose its brilliance and the atmosphere showed a great array of colors. The sun grew pale; the atmosphere became dull yellow; a light cloud very beautiful in form and bright appeared above the tree of the apparitions, remained there a few minutes, and finally rose and rose toward the heavens where it disappeared. Everyone's face glowed, rose, red, blue, all the colors of the rainbow. The trees seemed to have no branches or leaves but were all covered with flowers; every leaf was a flower. The ground was in little squares, each one a different color. Their clothes seemed to be transformed also into the colors of the rainbow. The two vigil lanterns hanging from the arch over the holy spot appeared to be of gold. The crowd was delighted. No one had seen the heavenly Lady, but these extraordinary phenomena, already observed at the previous apparitions, showed clearly that Our Lady had come to the meeting place.

Even though the children were not there, it was half past one when a column of smoke rose precisely from the spot where the children had been in the previous visitations. It was a thin, feeble, bluish column that rose straight up, two meters perhaps, above the heads of the people, and then broke up. This phenomenon, clearly visible to the naked eye, lasted a few seconds. The smoke broke up quite quickly and after a while the phenomenon repeated itself a second time and a third time. The tall columns were clearly seen in the gray atmosphere three times, principally the last.

The children did not see the beautiful lady on the August 13, but on August 15 in the afternoon the children were with the sheep in the pasture and felt something supernatural approaching. They saw a flash of light and then Our Lady appeared over a holmoak tree. Lucia asked, "What do you want of me?" She answered, "I want you to continue going to the Cova da

Iria on the 13th, and to continue praying the Rosary every day. In the last month, I will perform a miracle so that all may believe." The dialogue continued and when finished the Lady rose up into the air surrounded by her globe floating off into space.

Fifth Apparition: the beautiful lady in the globe radiates a stream of light covering Pluto's planet

From the previous evening (September 12) a witness wrote, "I saw the endless stream of people coming from a distance on foot in order to see the apparition; on the following day the crowd gathered at Fatima at about thirty thousand. I was deeply moved, and more than once tears came to my eyes on seeing the piety, the prayers and ardent faith of the many thousands of pilgrims who recited the Rosary on the way. There was not a road or pathway that was not full of people, and never before in all my life had I witnessed such a glorious and moving manifestation of faith."

Several times, the sun lost its brightness as some clouds covered it, and the clouds became in turn white, red, and yellow. These same hues were reflected in the faces of all present. Everyone was convinced that something extraordinary was happening. At noon, though not a single cloud was to be seen, the glorious sun of the radiant day began to lose its brilliance to such a degree that the stars were visible! The atmosphere became a golden yellow. Distinctly many saw a globe of light advancing from east to west, gliding slowly and majestically through the air. Then suddenly the globe with its wonderful light disappeared from sight.

At this moment, a light cloud was seen to form about the tree. A white cloud then hid the top of the oak, and at times the whole tree, together with the children. Lucia again spoke to the beautiful Lady but no one heard anything.

(After the globe ascended) Oh, Miracle!—from the limpid and cloudless sky there began to rain upon those present a shower, as it were, of white flowers, which on coming near the Earth vanished without touching it. Many rose petals fell. They came from the sun in large quantities. When high in the air they were large but on approaching the people they became small and vanished. Men held out their hats to collect them, but when they tried to hold them they found nothing.

Everyone looked at the luminous globe, until it disappeared in the direction the planet Pluto.

Sixth Apparition: the energy of this luminous globe empowers Pluto as never before

Our Lady had promised Lucia a miracle on the last apparition, October 13. From the eve, October 12, all the roads to Fatima were already packed with carriages, bicycles, and an immense crowd of pilgrims reciting the Rosary and singing hymns, on their way to the site of the apparitions, where they were going to spend the night in the open. It might be called a general mobilization to go and hear the message from heaven, and to see the promised miracle that was to authenticate the message. Though no one knew in what the miracle would consist, each was determined to see it at firsthand.

The morning of October 13 was disappointing, for contrary to all expectations it was wet, gloomy, and cold. It seemed as if heaven wished to test the faith and devotion of the pilgrims, and to make them merit, by a hard sacrifice, the honor of witnessing the promised miracle. But the bad weather did not check in any way the crowd that gathered even from the frontiers of the country, while reporters and photographers from the big newspapers were there to get the facts.

The continuous rain had transformed the place of the apparitions, which is a hollow, into a vast mud pit, and all, pilgrims or curious, were drenched to the skin or frozen with the cold. Shortly before midday, an observer estimated the crowd at about seventy thousand.

Suddenly, Lucia gave a slight start, and cried: "There is the lightning." Then raising her hands she added: "See, She is coming! She comes! Do you see Her? " Lucia no longer heard; she was in ecstasy. The clouds of smoke vanished as the luminous globe ascended.

Immediately the clouds opened wide, exposing an immense surface of blue. In the cloudless area the sun appeared at its zenith, but with a strange aspect. For though not a single cloud veiled it, while being very bright, it was not dazzling, and you could look straight at it at will. Everybody looked in surprise at this new kind of eclipse.

Suddenly the sun trembled, was shaken, made some abrupt movements, and finally began to turn giddily on itself like a wheel of fire, casting in all directions, like an enormous lamp, great beams of light. These beams were in turn green, red, blue, violet, etc., and colored in a most fantastic manner the clouds, trees, rocks, the Earth, the clothes and faces of this immense crowd, which extended as far as the eye could see.

214

After about four minutes the sun stopped. A moment later it resumed a second time its fantastic motion and its fairy-like dance of light and color such as could never be imagined in the most gorgeous display of fireworks. Once more, after a few minutes, the sun stopped its prodigious dance, as it to give the spectators a rest. After a short stop and for the third time, as if to give them an opportunity of examining the facts carefully, the sun took up again more varied and colorful than ever, its fantastic display of fireworks without a doubt the most glorious and most moving that had ever been seen on this Earth.

All through these unforgettable twelve minutes, during which this unique and gripping spectacle lasted, the enormous multitude was there in suspense, immovable, almost in ecstasy, breathless, contemplating this moving drama, which was seen distinctly within a radius of more than twenty-five miles.

It was the terrible fall of the sun that was the culminating point of the great miracle, the most awful moment and the most divinely moving. Indeed in the midst of its crazy dance of fire and colors, like a gigantic wheel which from spinning has swung off its axis, so now the sun left it place in the firmament, and falling from side to side, plunged zigzagging upon the crowd below, sending out a heat increasingly intense.

And with one accord, falling on their knees in the mud the spectators recited in a voice choking with sobs the most sincere act of contrition that had ever come from their hearts. Finally, stopping short in its vertiginous fall, the sun climbed back to its place, zigzagging as it had come down, and ended by gradually regaining its usual brilliance set in a limpid sky.

Moving detail: this apocalyptic scene full of majesty and terror ended with a delicate gift, which shows the motherly tenderness of the Heart of this beautiful Lady for Her children. Through all had been drenched to the skin, each now had the pleasant surprise of feeling quite comfortable, their clothes being absolutely dry, and the puddles of water on the ground had also dried up.

All of this radiated from the Earth to Pluto and it experienced all of this radiating back the purity and purification that came to it from all these people and this last apparition. The luminous globe with the beautiful Lady floated by Pluto one last time, a time to remember in Pluto's pure memory bank. This event marked a turning point in Pluto's evolutionary development, one that it will be marked with joy and purity forever more.

And this myth, Our Lady of Pluto, is a true story, just recounted as a myth that lives on in the heart of Pluto and those who contemplate its existence.

12

Love Song for Pluto

SHELLI JANKOWSKI-SMITH

I am sure you are familiar with the myth of Persephone, who
was forcibly abducted by Pluto and snatched away into the
underworld. Symbolically, Pluto describes a forced initiation which appears to come out of nowhere but changes our
lives forever. It is where we are taken over by forces we never
believed we were going to get involved with ... It is interesting
that there is an ongoing debate about whether Pluto is in fact
a planet ... from a psychological perspective, if Pluto does not
belong in the solar system then it doesn't belong within the
human psyche, and we don't have to take responsibility for it as
an aspect of ourselves.

From an interview with Clare Martin, author of "Mapping the Psyche:
An Introduction to Psychological Astrology" (posted on www.astro.com).

You, tough-love planetary swarm of moonlets, we can't even decide
what in God's name to call you. Swung out on the thread of some crazy
ellipse you warble there at the fringes of a universe we claim to know. And
as we move ever inward through time to understand ourselves, out there
you wait for us in the peripheral places of knowing. It's taken so long to
circumvent you. We love you, oh our distraction. For we love and fear the
things we seek in order to avoid scraping down to our own deepest crust.
So. We've put this off as long as we can. Here now, at last, is the inevitable
moment to rush toward everything that we've abandoned. Here now, dark
orphan, we will start a new dance with you. Can we finally be strong enough
to hold your cold heart against ours as together we wheel out painfully
through ecstatic space?

13

I Feel Bad about Pluto

Lisa Rappoport

How must it feel,
the only planet ever to be
demoted? Eternal chagrin,
cold as ice. Pluto—forever a loner,
emblematic of what is far,
the jumping-off place
where one departs
the tiny known for the vast
unknown, our ninth, our final ring—
is now a mere lurker, a frozen rock
pushing in where it's no longer
wanted. What can you expect
from a celestial body named
for the god of the underworld?
Once it defined the edge
beyond which were dragons; now
it *is* the dragon, slain. Like an ex-
lover should, it has lost all allure. Rotating sadly
in exile, this dwarf planet's status
was besmirched for lack of "clearing
the neighborhood." By swallowing
or repelling smaller objects in its orbit,
consumption or rejection, Pluto
would redeem itself. And it may
still, though who will be here to take note,
and in what languages, and from what
vantage point in the cosmos, remain

chilly questions sent out into the galaxy
to travel, perhaps endlessly,
at the speed of light.

14

Pluto

ROBERT KELLY

I never said I was symmetrical, I'm more
like a cripple, I wobble and have no moons
to call my own though some have said I do
to me from time to time and I have bedded
them deep in the distance of your house.
Yes, I am the edge of you, I am as far
as your famous Sun knows how to shine
if that and I am always cold. Lately
they tell me I am not even what I thought
for eighty years I was, a planet, small
but on the list of members. And now not.
How would you like it if you went
to the tax man to pay your taxes and she
said No, we won't take your taxes, you
are not a citizen, not even a person
and only persons are taxed in this land
so far. Wouldn't you be horrified, yearn
to ante up your 28 percent like other people
and belong? But I'm not people.
For millions of years I've been
the black sheep, the invisible influence,
blamed for catastrophes, the scourge
of astrologers, the lord of exile,
named for the Invisible God who ruled
the shadowy land where the dead
in their terrible paradox were
understood to live. For some

reason they thought I was rich,
maybe they imagine all the
Stuff You Can't Take With You
comes up here to me. Up, I say,
but it is truly down. I am below
anything you understand, when
you look in your telescope it's not
me you see but my shadow in the sky
and the shadow of darkness is light.
Now I am an exile, exiled even
from my own name, am given
some trumpery number as if I were
just some pompous asteroid,
those gypsy vagabonds whose ice
reflects emotions back to earth
but you don't know that yet.
You poor fools. I am the lord
of limits and the queen of edge,
I am experiment and failure,
glorious failure and long life,
I am artifice and substitute,
I am the other side of nature
and if you're lucky even you
will get to me some day.

5 February 2014

15

Pluto

MAGGIE DIETZ

Don't feel small. We all have
been demoted. Go on being

moon or rock or orb, buoyant
and distant, smallest craft ball

at Vanevenhoven's Hardware
spray-painted purple or day-glow

orange for a child's elliptical vision
of fish line, cardboard and foam.

No spacecraft has touched you,
no flesh met the luster of your

heavenly body. Little cold one, blow
your horn. No matter what you are

planet, and something other than
planet, ancient but not "classical,"

the controversy over what to call you
light-hours from your ears. On Earth

we tend to nurture the diminutive,
root for the diminished. None

of your neighbors knows your name.
Nothing has changed. If Charon's

not your moon, who cares? She
remains unmoved, your companion.

16

Falling in Love with a Plutonian

Dinesh Raghavendra

Dreaming, crumpling paper balls at
An unkempt desk, hoping the origami
Slips an idea in my head.
Endless days and nights wondering
About the celestial swarm. I weave
A story: fall in love with a plutonian
Meet her as Charon ferries you
Across the Styx. Kiss the girl
Who has a missing mouth
She's only animal from the neck down

She makes art from her bodily
Secretions. She will penetrate you
Like an insect. Intersect your nightmares
With her dreams. Inhabit your id.
Turn you into an interstellar archetype.

A decade later, a probe finds you
lost in the wheels, the cogs, the gears;
Of the process turning in your head
Wandering the deserts of Cerberus
With only her for company.
Fall in love with a Plutonian.
Save the planet with her and a satellite
Sit with her by the rivers of Hydra
As it dries to a trickle in the sand
Watch the stars.

Falling in Love with a Plutonian

You are her birdman in a skydungeon
With dreams of astronomy
And nightmares of astrology
Planet after planet of almanac squares
The woman-smell
in the small planet
down there.

17

Dostoevsky's Pluto

STEVE LUTTRELL

"As above, so below"
and out of the elements,
one of the pagan trinity,
with Zeus and then Poseidon,
Pluto stands.
The sovereign of the underworld,
ruler of the dead.
A lonely lord
eternal with his thoughts
and in chthonic darkness
his sad, abducted bride,
a captive's measure keeps.
As in this place
a wordless history's kept
and this is, as the author says,
Notes from Underground.

18

Ten Things I'd Like to Find on Pluto

PHILIP WOHLSTETTER

1) Graffiti.

2) A black monolith.

3) The song of the Sirens.

4) Amelia Earhart's plane.

5) The time that Proust lost.

6) An outside to capitalism.

7) Philip K. Dick in conversation with Chris Marker.

8) A used condom.

9) Something I couldn't possibly imagine.

10) My mother and father in their best moment.

What do I think the probe will find?
None of the above.
What would change some key paradigm or meaning?
All of the above.

19

Plutonic Horizons, or
My Sixty-Nine-Year Search for Planet X

PHILIP WOHLSTETTER

There were nine planets in my solar system. I expected to visit them all. I memorized their names in grade school, tracked their orbits along the ceiling of the Copernican room in the Hayden Planetarium where replicas of the planets revolved around a replica of the Sun; or rather six of them did, because the orbits of the other three if built to scale, according to a lecturer in popular astronomy, would put them outside of the building, and distant Pluto, discovered a mere fifteen years before I was born, would be inscribing its long, lazy ellipse between the north end of the Central Park reservoir and the Automat on Forty-Sixth and Broadway. My two favorite books were *The Conquest of Space* and *Across the Space Frontier,* not so much for the text by (among others) Willy Ley and Wernher Von Braun, though that wasn't bad, but for the illustrations by Chesley Bonestell, an architectural designer and Hollywood special effects artist whose close-up renderings of remote planetscapes were both precise and breathtaking. Saturn was his specialty—Saturn as a ringed, spectral presence, visible only in outline in the sky above Titan; Saturn as an enormous spinning ball looming, yellow and lopsided, over the rim of Mimas. He imagined the surface of Pluto, its serrated ridges, spreading fields of ice, and above it, he depicted our Sun in the vast blackness of space, slightly larger and brighter than other stars but not at all conspicuous. If *Life Magazine* or *Colliers* laid out the military or economic argument for space travel, it was Bonestell who incited the delirious interplanetary desire that Kubrick would revive a decade later.

Space travel in those days was the jewel in the crown of the Future—the sexy part of it—and the Future was still thriving then. Disney had made it into a franchise called Tomorrowland. General Electric kept advertising its immanent arrival with the slogan, "Progress is our most important product."

Robots to do the shit work, colonies on the moon, orbiting hotels with grand Cinemascopic views, flying cars and jetpacks for navigating through multi-level rush hour traffic back on Earth: what was not to like? Guaranteeing the bright panoramas awaiting us, authorizing the show from behind the curtain so to speak, was an invisible wizard, a tall Englishman with a mustache by the name of John Maynard Keynes. In a 1930 essay called "Economic Possibilities for our Grandchildren," the world's most famous economist had peered into his crystal ball to discern the outlines of life on Earth in 2030.

He saw an Age of Abundance ahead. Automation will have reduced what Marx called "the working day" to three hours. The "absolute needs" of food, clothing, and shelter will have been met. As for merely "relative needs" like yachts or high status toys, needs driven by the desire to feel the equal or the superior to others—by what Rousseau would call envy, Rene Girard (in a more complex key) mimetic desire, and 1950s American sociologists "keeping up with the Joneses"—these illusory needs, though seemingly limitless, will slowly lose their power to compel people to work when the possibilities of more free time become clear. "For the first time since his creation man will be faced with his real, his permanent problem—how to use his freedom from pressing economic cares, how to occupy the leisure which science and compound interest will have won for him." True, Keynes never imagined the exploration of space as a solution to the problem of leisure, but space opera SF often presupposes a Keynesian post-scarcity world. Social theorist Peter Frase noted this while watching *Star Trek: the Next Generation*. The technological conditions for Star Trek World are (1) a replicator "that can make instant copies of any object with no input of human labor," and (2) an unlimited source of free energy via the magic of anti-matter or dilithium crystals. "There is no money, everyone has access to whatever resources they need, and no one is required to work. Liberated from the need to engage in wage labor for survival, people are free to get in spaceships and go flying around the galaxy for edification and adventure." (Interestingly, Frase calls this post-capitalist society "Communism.")

"The Real," says Slavoj Zizek, "is the disavowed X on account of which our vision of reality is anamorphically distorted." Pluto was first referred to as Planet X, a hypothetical, an absent presence suggested by perturbations in the orbit of Uranus not attributable to Neptune. X indicates the unknown in an algebraic equation. X marks the spot on a treasure map or the files no one may see. X can refer to that which exceeds a situation or that which structures one according to a principle incompletely understood. Sometimes I frame my life's journey as an unconscious quest for an X factor, the invisible planet that threw the Future off course. I wasn't aware of any such X on July 20, 1969 when Neil Armstrong stepped down from the lunar module onto the surface of the moon. Sure, I'd seen the visible cracks in the Future. Vietnam. Alabama. Newark. Watts. But standing before a TV in Burlington, Vermont with two fellow actors from the Champlain Shakespeare Festival and a group of kids from Harlem sent up to the Green Mountain state for an outdoor experience, watching the tiny figure in a spacesuit work his way down the ladder onto the Sea of Tranquility, the interplanetary future seemed big enough for all of us. For all mankind, in fact.

Still, there were two people, a Briton and an American, the best SF writers of their era (or perhaps any), who understood at the time that the Age of Space Travel was already over, that the Promised Future would never arrive.

Chapter Three of *The Three Stigmata of Palmer Eldritch* is Philip K. Dick's bleakest vision of settler society, a Mars outpost well on its way to becoming the next Lost Colony of Roanoke. Driven from Earth by the onset of global warming, the settlers live in hovels, listless, troglodytes in a dreary landscape waiting for resupply by a monthly UN space shuttle. No one works any more. They do a lot of drugs. The hydro-pumping system is in disrepair, irrigation ditches are filled with sand, and microscopic native pests shred the leaves of abandoned gardens. Favorite pastime: playing Ken-and-Barbie with dolls named Walt and Pat in a layout of a furnished world. The drug of choice, an illegal substance called Can-D, enables them to literally inhabit their Walt and Pat dolls and experience "the near sacred moment in which the miniature artifacts of the layout no longer merely represented Earth but became Earth." Somehow they project themselves into Perky Pat's sunny, futuristic San Francisco: they can jet around in a Jaguar XXB flying car or slip into bathing suits and picnic on the beach, untroubled by the future. ("It's going to be hotter ... But we won't live that long.") Ontological uncertainty persists, however, even in the giddy whirl of pleasure. Are the game-players in a drug-induced hallucination, they wonder, or have they really been translated somehow from Mars to Earth?

In his brilliant story, "The Cage of Sand," J. G. Ballard actually brings Mars back to Earth, millions of tons of Martian topsoil dumped as ballast on the coast near Cape Canaveral because the much-lowered gravitational mass produced by the continuous launching of rockets and space probes threatens, if not remedied, to alter the Earth's orbit and send it closer to the Sun, resulting in higher temperatures, loss of the ozone layer, and destruction of the biosphere. Unfortunately, the Martian topsoil contains a virus deadly to plant life. ("All too soon it was found that the microbiological analysis of the sand had been inadequate.") So much for geo-engineering. In the aftermath, Cape Canaveral is a ruin. Around it are ghost towns. Abandoned motels. Submerged chalets. Intact cocktail lounges with names like *The Orbit Room* or *The Satellite Bar,* their half-occluded signs stuck in the red sand amid rusting fragments of Mars rockets. Living in this landscape of disaster where old launching gantries and landing ramps rear up into the sky "like derelict pieces of giant sculpture," evading the Wardens who patrol the fenced-in area to enforce a quarantine, are a typical group of Ballardian obsessives drawn to a catastrophe that mirrors their inner lives. Louise Woodward has returned to the last place she saw her husband alive. An astronaut whose vessel failed to make contact with a space platform, his body orbits the Earth in its capsule, periodically intersecting the orbiting capsules of six other dead astronauts, Russian and American, killed on other failed missions over the past fifty years. Every night, she keeps her vigil, waiting for her husband to pass overhead. Travis is a failed astronaut. He lost his nerve during the countdown "lying flat on his back on a contour couch two hundred feet above the launching pad," costing the aerospace company five million dollars. "Night after night, he watched the brilliant funerary convoy weave its gilded pathway toward the dawn sun, salving his own failure by identifying it with the greater but blameless failure of the seven astronauts."

Certainly global warming is a huge X looming over us. That alone would make these stories relevant but it's not what makes them resonate for me. It's the way they transform the indefinitely postponed future into the uncanny. Failure is the default mode in both Ballard and Dick. They'd seen it on TV every night in the Vietnam War, despite the can-do attitude of the generals, and they imagined it spreading its stain over things to come. Ballard even inscribes it into nature by having the funerary capsules become celestial objects "shining with the intensity of second-magnitude stars." Each new object is given the name of the dead astronaut inside it—Connolly, Tkachev, Woodward, Maiakovski—part of an "orbiting Zodiac," a new constellation.

"They were disposed in a distinctive but unusual pattern resembling the Greek letter x . . ." Ah, that X again! I can even locate a metaphorical X in Dick's novel, the planet Pluto itself, a threshold between the human and the alien. From out of a spaceship crash landing on the surface of the planet emerges Palmer Eldritch, intergalactic drug dealer / entrepreneur. Eldritch (or his alien simulacrum) has returned from Proxima Centauri bringing with him a new hyper-addictive drug. Unlike the Can-D ingested by Mars colonists, which allows them to share the pleasant experience of Perky Pat's world, Chew-Z imprisons each user in his or her own solipsistic world. ("There's no such thing as society," a British prime minister might think if she was using it, the neo-liberal drug of choice.) Factor in that these myriad worlds are all controlled by Palmer Eldritch and you're left with another model of an outside X not immediately perceivable inside the world it structures.

In 1930, eleven days after the celestial X finally revealed itself to Clyde Tombaugh of the Lowell Observatory in Flagstaff, Arizona, the new planet was given the name *Pluto*. Some etymology. Πλουτος (ploutos) is the Greek word for wealth. As a proper noun (ΠΛΟΥΤΟΣ), it refers to the Greek god of wealth, Plutus (not to be confused with Pluto, the Roman God of the underworld). From πλουτος come words like *plutocracy* (rule by the wealthy) and *plutonomy* (defined in a helpful report from Citigroup as an economy in which there is no average consumer, only the rich and the rest). Wealth, I have come to believe, is the Planet X I've been searching for, the invisible object whose gravitational pull causes the perturbations in the orbits of our lives, but for reasons which I'll clarify when we examine why Keynes's vision of life in 2030 was so out of whack, why that Radiant Future beamed into my head in the '50s—that future of leisure and abundance and space travel for all—seems to have been cancelled, I want to call this Planet X "Capital."

Keynes's essay reads now like a work of fantastic literature. Three-hour day? Fifteen-hour workweek? According to a Gallup Poll (Sept. 2, 2014, *Washington Post*), an average workweek in the United States is 46.7 hours with 21 percent of workers clocking in 50-59 per and 18 percent working 60 plus hours (only 13 percent of workers like their jobs). Wealth for all through the "miracle of compound interest"? Tell that to students, payday loan customers, and underwater homeowners collectively compounding a debt that can never be repaid. Here are the prices—the astronomical prices—paid by high-net-worth individuals in software, telecom, and finance to hitch a ride on a Soyuz rocket out to the Russian space station: Dennis Tito, twenty million dollars (2001); Mark Shuttleworth, twenty million dollars (2002); Gregory Olson, twenty million dollars (2005); Anousheh Ansari, twenty million dollars (2006); Charles Simonyi, twenty million dollars (2007), thirty-five million dollars for a second trip (2009); Richard Garriot, thirty million dollars (2008). My imp of the perverse wants to connect these numbers to a Keynesian sentence heavy with the ruins of the Future: "They (the wealthy classes) are, so to speak, our advance guard—those who are spying out the promised land for the rest of us and pitching their camp there." No author should be booted from the canon for failing to predict the future; otherwise Marx wouldn't be essential reading. What Keynes failed to see was the present. He pulled off the incredible feat of saving capitalism without, in fact, knowing what it is he saved. To understand why Capital always promises a future it can never deliver, a return to *Star Trek* is necessary.

Recall that sociologist Peter Frase pointed out how *Star Trek* is set in a post-scarcity world with advanced technology and renewable energy. Economics, often defined as the allocation of scarce resources among competing ends, has no place here. Nor does wage labor, private enterprise, or the pursuit of profit since the Replicator can make an infinite number of copies of anything one could possibly want. Frase then poses the question: How could Star Trek World be transformed into a capitalist society? His answer: intellectual property. Require a license to be bought for every product created and immediately there is a need for lawyers to draw up licenses, ways to earn money to buy the licenses, guard labor to work in prisons to confine violators of the terms of use and so on, until pretty soon competition and profit are driving the system. In short, create artificial scarcity. What's the point of this thought experiment? If the fictions of a State of Nature and the Social Contract illustrate the liberal view of a political society grounded in an agreement between individuals, Frase's leftist fable of the Anti-Star Trek Society reveals in an elegant way the core of capitalism. Take away scarcity, production, exchange, and you're still left with Capital's interminable drive to expand, its need to reconfigure whatever world it finds itself in so it can continue to valorize. This is what Keynes totally misses. Underpinning his vision of the future, as noted, is a distinction between absolute needs, which the economy can satisfy, and relative needs, which are limitless but insubstantial. But there is a limitless need that keeps the system going and it's not a human need.

The question Keynes should have asked is: What does Capital need? It wasn't as if hordes of deadbeat poor people were swarming into banks begging for sub-prime loans, as the Tea Party tells the story; it's that bankers were cold-calling everyone in the phone directory because of the never-ending market imperative to keep returns rolling in, and fees from guaranteed-to-fail loans generated higher ones than traditional banking. Money always wants more money (just like information wants to be free). What Capital needs to do is to keep reproducing itself, to keep the ball rolling. If it needs to sell you the same things again and again, cars every few years, DVDs in the Special Edition, the Director's Cut, the Blu-Ray, the superfragilistic Xeno Ray, it will find a way to do so. If clothing sales are peaking, it will invent Brands. If water is a public resource, it will privatize it and sell it back to you. This need to keep expanding is inscribed within the system. Keynes's mistake was to reduce an objective economic structure to an effect of psychology. The circuit of Capital is like a treadmill that will never stop on its own, a parody of Nietzsche's eternal return, a scratched LP that just keeps repeating the same notes, M-C-M (Marx's shorthand for Money—Commodity—Money), and the promise of a future of abundance and cool new gadgets for everyone is just another ingredient that keeps the machine running, a lure for versions of the ten-year-old me in the '50s, poring through illustrated SF magazines and haunting planetariums.

This year, a fragment of that lost future known as the past has floated into my consciousness again. On July 14–15, 2015, an unmanned NASA space probe will fly by the surface of Pluto. Launched on January 19 in 2006, it will have taken nine and a half years to reach the outer edge of our solar system, the long slow time of interplanetary travel. Mission time is not the time of our lives, nor the frantic, hopped-up time of the reproduction of Capital along its circuits. *New Horizons*, as this voyage of discovery is called, is the third attempt to mount a mission to Pluto. The first two efforts, *Pluto Fast Flyby* and *Pluto Kuiper Express,* were aborted because of budget cuts. In the end, space travel is a mirror of the present. While astronomer Neil deGrasse Tyson appeals to Congress for an infinitesimal raise in funds for NASA, libertarian millionaires hop rides to the Russian Orbital Station, another bit of

Soviet state property auctioned off to ex-apparatchiks in the great neoliberal fire sale. Maybe these bozos need an escape route off a planet their enterprises have trashed. Or maybe there'll be a mass exodus and our solar neighborhood will end up resembling some demented version of a Kim Stanley Robinson novel. Maybe every moon, asteroid, meteor, ring, and plutino capable of being terraformed into a habitable berth by a brave new technology will become a city, cabana, or time-share, an apodment in some extraterrestrial Tokyo or Manhattan. Or maybe not. Maybe what Alex Williams and Nick Srnicek call "the promissory note of the mid-Twentieth Century's space programmes" will turn out to be redeemable when all bank-issued debt is worthless. Maybe a new wave of Left techno-social imagining can reinvent space travel as a form of aesthetic play, an energetics of eternal delight to take the place of work. History is full of surprises, another way of saying that the letter X will still haunt us.

So what will the probe find on Pluto? A black monolith? The ruins of Palmer Eldritch's crashed ship? An outside to Capital? I'll be curious to see. Since anything I can name is not really an X, I'm hoping for something else—something not only stranger than I imagine, but stranger than I can imagine.

20

Ten Things I'd Like to Find on Pluto

Jonathan Lethem

1) The fourth spatial dimension.
2) The world before Homo sapiens.
3) The world before the Internet.
4) A Robert Smithson.
5) All my dead pets, travelling together in a cosmic version of *The Incredible Journey*.
6) A complete cut of *The Magnificent Ambersons*.
7) Change we can believe in.
8) A genuine anarchist society.
9) A genuine communist society.
10) That which cannot be named but for which we all yearn.

What do I think the probe will find?
Only what it has been told to look for.
What would change some key paradigm or meaning?
If Pluto found us before we found Pluto.

21

Ten Things I'd Like to Find on Pluto

ROBERT SARDELLO

1) Pure Stillness
2) Pure Silence
3) Clear Vision
4) Luminous Darkness
5) Soul Perspective
6) Soul Wealth
7) Clear Reflection
8) Deep Dreaming
9) The Interior of Matter
10) New Forms of Love

What do you think the probe will find?
Utter Stillness animating the Soul.
What would change some key paradigm or meaning?
Love-filled Matter.

22

Ten Things I'd Like to Find on Pluto

ROSS HAMILTON

In Western alchemy, it is the Chaos or Plutonian element that assists the prac-
titioner in solving the mystery of death (putrefaction). That element affects
the true maturation of sulfur—the red brimstone of our wisdom without
which we wouldn't bother cultivating the elite science. Red sulfur alone is
a fine medicine, but ultimately we must progress through the eight (or nine)
stages of Mercury's magic mirror to acquire the lights of complete maturity.
So in a sense, it is between Mercury and Pluto that the whole idea or theory
is properly grasped. This system originated in India and China, but its adaptations to European thought have intrigued poets and scholars for centuries.

Like the Pluto assemblage out at the edge of our solar system, this agent
works quickly upon the peripheries of certain of the elements and also certain receptive molecules containing those elements, eventually penetrating
and softening them all with Eros—the mysterious aspect of the eternal ether
(Sanskrit: *Akash*) that pours from it as flux pours liquid from a magnet. It is
a *universal,* and the change it brings joins and harmonizes. It is in this one of
the best kept secrets of our divine art. Similarly, Pluto's orbit does not conform to the other orbits of the planets, and so it is that angle of eccentricity coupled with its penetrating drive that enables its sacred, encompassing
journey of ingress with the other orbs' orbits about our Sun:

> We watch the swarming Pluto as he winds his crafted way—
> To circumnavigate the altar of his siblings, father Sol
> On an elliptic lane—not the ecliptic plane
> As the others all conforming dutifully must roll.
> We watch the swarm of Pluto as the alchemist of fate
> Tips his patient hand to shine a brighter light the laws
> Of physical phenomena—a glimpse into our competent
> Industry initiate philosophers give pause.

Ten Things I'd Like to Find on Pluto

Our solar system is one of countless, such that were conceived and brought about long ago by a magical process we do not understand. We are in the glass vessel of the Creator. A lifetime study of all the mysteries including geometry, mathematics, alchemy and its chemistries, telescope, and microscope, may one day illumine the adages "As above and within, so below and without," and "Soul, know yourself."

I'd like to suggest a top ten of possible things we may find happening *as a result* of our finally arriving at the Pluto group:

1. We may find that just as *Harry Potter and the Philosopher's Stone* was changed to *Harry Potter and the Sorcerer's Stone* for a dumbed-downed American audience, no one will really care about the Pluto probe's findings unless the coverage is tweaked down a tad here in the States. Maybe start with the rag sheets and ever so gradually work up to No' Adams.

2. Really though, when we finally are in propinquity to the Pluto pride, we'll recognize debating is over and new data gathering is just beginning. This results in the reopening of the debate around the family dinner table in Brooklyn when Jr. declares, "Hey Mom, I just found out that Pluto's more like a binary entity and only one in a swarm of kindred objects—go figure!" Dad says, "Huh?"

3. Nassim Haramein's suggestion that the planets are not rotating about our Sun but instead are in a swarming corkscrew tagalong as old Sol speeds through space will prove to have greater clarity from the angle of Pluto.

    It may take a swarm to see a swarm.

4. We may find evidence to support our Pluto having secretly been the initial inspiration for Sitchin's Nibiru (Lowell's Planet X), in spite of him working like a master sculptor to adapt and stretch the feasibility of an object considerably surpassing Pluto's 248-year transit about our star. Nibiru must be cold indeed, explaining the heartless treatment of the hapless Igigi. Tsk, tsk.

5. It is not impossible we may find that the six objects of the Pluto swarm, viz. Pluto, Kerberos, Nix, Charon, Styx, and Hydra, are actually worthy of being reclassified as small planetoids, each of which themselves has five moons, with each of them in turn having five sub-lunar spheres, and each of them in turn five, and so on. This finding results in the probe

making a concerted effort to map the first interstitial space fractal—the detailing creating a feedback loop that shuts its communications systems down.

6. Not believing the fractal story behind the shutdown, some scientists may find that the swarm is in realty a sort of disparate black dwarf in the final stages of collapse. This creates concern for the future of neighboring Neptune.

7. Not believing either of the above, another group of scientific observers seizes the opportunity to lay blame on incompetency, making the case that we should have been prudent, waited longer, saved our money, and then abandon the whole idea in favor of a bigger space telescope.

8. The long held belief that Pluto & co. is *so small*—especially as it follows the gas and ice giants of Jupiter, Saturn, Uranus, and Neptune away from our Sun—becomes compromised in lieu of the discovery that the region is filled with an equally giant, nearly undetectable transparent quartz sphere owning a small frigid core of rotating parts mimicking Vatican City.

9. We may finally put an end to the debate over whether an Irishman, Scot, or Russian can exist anywhere near Pluto when it becomes clear that the climate is inhospitable for growing barley, wheat, hops, and even potatoes.

10. We may discover that, having made the effort to explore every major object in our solar system, a friendly alien species of humanoid form will deem the occasion opportune for first contact. They will not share their advancements in technology however, instead accepting speaking engagements at key centers of learning to lecture on how far we still have to go before we are civilized enough to *stop mixing Greek and Roman deities* in the nomenclature of solar system objects.

What do I think the probe will find?

A bit more than our scientists expect. But *unlike* the finding of fresh marrow tissue within the petrified bones of T. rexes in Montana a few years ago (as shown on *60 Minutes*), they'll be able to explain most everything without any monstrous gaps in their knowledge base.

What would I like it to find?

A telling key to reveal the living nature of galaxies through the cumulative understanding that our own was designed and encoded to a small spermatic seed before release in the *Big Orgasm* along with uncountable others. I would like to find that, like some exquisitely planned and flawlessly executed pyrotechnical singularity, the Big O will be realized in time as a creation intended to prepare and populate virtually infinite space-time in the physical realm with conscious life in unending variety.

What would change some key paradigm or meaning?

Just the fact that we actually completed researching all the largest known objects in our solar system will change the model and the meaning of space exploration. It's about species maturation through the fulfillment of desire born of practical curiosity—and in a perfect world clears the air for more advanced and peaceful pursuits that we can all share in. But honestly, this isn't a perfect world, so how can we continue to foot the bill for the enormous expenses while still making real and timely progress?

East Indian masters first popularly introduced to the West the inborn ability we all possess to see and explore the cosmos internally with a simple reversal of the attention. We may access a faculty that is conveniently connected to everything—and it's basically free of charge.

Notes

1. Nassim Haramein and The Resonance Project Foundation, "Solar System's Motion Through Space," YouTube video, posted by Jamie Janover, December 5, 2009, https://www.youtube.com/watch?v=zBlAGGzup48.

2. Lesley Stahl, "B-Rex," *60 Minutes*, November 15, 2009, YouTube video, posted by fossildna, July 29, 2011, https://www.youtube.com/watch?v=2mDo8k-mtUM.

23

What the Probe Will Find, What I'd Like It to Find

JEFFREY A. HOFFMAN

What do you think the probe will find?

Assuming everything works OK, we will get the first photos of the surface of Pluto and its satellites. Also measurements of a thin residual atmosphere, if it hasn't already frozen out.

What would you like it to find?

Something surprising, as has been the case for first images of all the other planets we have explored.

What might it find that would change some key paradigm or meaning?

Given the confusion on Pluto's status as a planet and the horrible job that the astronomical community did in communicating this to the public, I'd love to see something that made Pluto a little more special than just being the "innermost Kuiper Belt Object."

24

Ten Things I'd Like to Find on Pluto

COLLEGE OF THE ATLANTIC STUDENTS

Participants: Joshua Noddin, David Cipollone, Joshua Sawyer, Sarah Rasmussen

1) DNA
2) Organic seeds from outside the solar system
3) Some never-before-seen quasi-organic self-replicator
4) Bacteria
5) The outpost of another species or an alien civilization
6) Evidence that aliens were shipwrecked on Pluto
7) The source of all the dark matter in the universe
8) That Pluto and its moons are each made of different kinds of cheese (including vegan cheese)
9) Giant fossils of life forms or of a civilization
10) Different physical patterns generated by Pluto's chaotic orbit and rotation leading to a different perception of space and time

What do I think the probe will find?
Probably nothing. Rocks and a lot of open space.

What would change some key paradigm or meaning?
Anything but rocks and space.

25

Ten Things I'd Like to Find on Pluto

NATHAN SCHWARTZ-SALANT

1) Persephone
2) The spiritual truth about wealth
3) The mystery of manifestation
4) Persephone
5) The secrets of dream consciousness
6) Birth from terror
7) The secret of healing the subtle body
8) Persephone
9) Penetrating vision
10) Why mothers exist

What do you think the probe will find?
Very dense, impenetrable matter.
What would change some key paradigm or meaning?
Revaluing disorder.

26

The Ten Worlds of Pluto

Charley B. Murphy

1. Planet of Women

Pluto is a totally female totalitarian society run by the masked Queen Illyana with an atmosphere amazingly similar to Earth's. The queen banished all males to a moon because they were responsible for the nuclear war that left her face burned, now hidden behind a mask. Her warriors dress like Rockettes, carry lethal rayguns, and live in the capital city of Kadia somewhat reminiscent of ancient Bagdad. Contact with Earth encourages the rebels led by Talleah to foment a revolution, eventually destroying Illyana's people and her deadly Beta Destructor aimed at Earth.

2. Anthromorphia Disnophinia

The Pluto mission finds a strange world populated by anthropomorphic "animals" (dogs, cats, mice, ducks) who wear clothes and speak English. Oddly, the citizens of the planet have mute "pets" which resemble hairless primates with features resembling Homo sapiens. They dress their pets in faux animal fur.

3. Cthulhu Calamari

A planet of hyper intelligent squid-like creatures inhabiting all ecological niches. Their leaders call themselves the "Old Ones." They have their own language and culture but understand English and have expressed an interest in exchanging diplomats with Earth, claiming it is their rogue colony. They are upset to hear that calamari is a popular food on Earth.

4. Genital Face Planet

This planet is inhabited by a race of beings similar to male and female (and intersexed) Homo sapiens except that their "faces" resemble human genitalia. It is determined that under their clothing they conceal organs resembling human faces that they cover up in public. These "lower faces" do not speak but function primarily for excretion and reproduction through variations on "French kissing."

5. Teenage Werewolf Vampire Planet

There are two dominant species on this planet, both with a relatively short lifespan. The vampires feed primarily on the blood of a specially bred humanoid population. The second dominant species, the werewolves, are primarily carnivorous (raiding the brood humanoid population of the blood-sucking species) but also hunting wild animals. In order to eat or fight they change at will into large canines (wolves). The two main species are periodically at war but are also capable of intermarriage. Successful interspecies mating can produce a hybrid subspecies with somewhat random characteristics of its parents. The dominant species cooperate in planetwide competitive games similar to our Olympics.

6. Mad Water World Maximillia

There is very little dry land on this ocean planet. Its great seas are filled with myriad fish, amphibian, and aquatic mammalian life. There is one humanoid species with rudimentary gills living on boats, subsisting primarily on fishing and practicing agriculture on floating debris islands of recycled organic matter. The few islands of dry land are inhabited by a warlike humanoid species scavenging in the dystopian remnants of a carbon energy-based industrial civilization. When the two species interact it is primarily via the pirating raids of the land dwellers on the mutant "gill people."

7. Fifty-Eight Genders Planet

The following genders have been catalogued on this planet: Agender, Androgyne, Androgynous, Bigender, Cis, Cisgender, Cis Female, Cis Male, Cis Man, Cis Woman, Cisgender Female, Cisgender Male, Cisgender Man, Cisgender Woman, Female to Male, FTM, Gender Fluid, Gender Nonconforming, Gender Questioning, Gender Variant, Genderqueer, Intersex, Male to Female, MTF, Neither, Neutrois, Non-binary, Other, Pangender, Trans,

Trans*, Trans Female, Trans* Female, Trans Male, Trans* Male, Trans Man, Trans* Man, Trans Person, Trans* Person, Trans Woman, Trans* Woman, Transfeminine, Transgender, Transgender Female, Transgender Male, Transgender Man, Transgender Person, Transgender Woman, Transmasculine, Transsexual, Transsexual Female, Transsexual Male, Transsexual Man, Transsexual Person, Transsexual Woman, Two-Spirit, Male, Female.

8. Gojira Kaiju Planet

A planet of large, warring creatures called Kaiju. A partial list includes: Godzilla, Anguirus, Snowman, Meganulon, Rodan, Moguera, Varan, Orochi, Vampire Plant, Mothra, Maguma, Giant Lizard, Oodako Giant Octopus, King Kong (Toho), Matango, Manda, Dogora, King Ghidorah, Frankenstein, Baragon, Gaira, Sanda, Ebirah, Ookondoru Giant Condor, Mechani-Kong, Gorosaurus, Giant Sea Serpent, Kamacuras Gimantis, Minilla Minya, Kumonga Spiega, Gabara, Maneater, Gezora, Ganimes Kamoebas, Hedorah, Gigan, Jet Jaguar, Megalon, Mechagodzilla, King Caesar, Titanosaurus, Shockirus, Biollante, Dorats, Godzillasaurus, Mecha-King Ghidorah, Battra, Godzilla Junior, Fire Rodan, Mechagodzilla 2 Super-Mechagodzilla, SpaceGodzilla, Kumasogami, Kaishin Muba, Amano Shiratori, Utsuno Ikusogami, Destoroyah, Fairy (Fairy Mothra), Garu, Desghidorah, Mothra Leo, Ghogo, Dagahra Dagarla, Barem, Rainbow Mothra, Aqua Mothra, Primitive Mothra, Armor Mothra, Eternal Mothra, Zilla, Baby Zillas, Orga, Meganula, Megaguirus, Kiryu, Monster X, Keizer Ghidora, "Mutos."

9. Ayn Rand Planet

A planet of large fortified "keeps" occupied by one person (or a nuclear family) living in high-tech anarchy characterized by sadomasochistic sexual encounters.

10. Oregon II

Similar to the movie "Another Earth" where a duplicate earth is discovered, except this Pluto duplicates only Oregon. Everything, including coffee shops, funny hats, bike paths, and great local brews.

27

Ten Things I'd Like to Find on Pluto

Timothy Morton

1) Something I had no idea I'd find on Pluto.
2) Disoriented stars wheeling.
3) Pluto's frozen surface is a gigantic convex mirror in which everything else appears distorted.
4) It is so cold that anything at all becomes quantum-smeared. Human eyes can see things shimmer while remaining still.
5) Things that look like puppets, suspended in solid nitrogen.
6) Care, tinged with a feeling of unease.
7) A secret that could not be revealed, only known as a secret.
8) A rock formation whose striated patterns uncannily resembled Yoko Ono's "This Is Not Here."
9) Longing to exit the solar system.
10) Longing to remain in the solar system.

What do I think the probe will find?

A small, dusty toy gorilla belonging to my daughter that she lost four years ago.

What would change some key paradigm or meaning?

The absence of life, in such a way that its absence remains uncertain.

28

The End of the World

TIMOTHY MORTON

Y ou are walking out of the supermarket. As you approach your car, a
stranger calls out, "Hey! Funny weather today!" With a due sense of
caution—is she a global warming denier or not?—you reply yes. There is a
slight hesitation. Is it because she is thinking of saying something about global
warming? In any case, the hesitation induced you to think of it. Congratu-
lations: you are living proof that you have entered the time of hyperobjects.
Why? You can no longer have a routine conversation about the weather with
a stranger. The presence of global warming looms over the conversation like
a shadow, introducing strange gaps. Or global warming is spoken—either
way the reality is strange.

A hyperobject has ruined the weather conversation, which functions as
part of a neutral screen that enables us to have a human drama in the foreground. In an age of global warming, there is no background, and thus there
is no foreground. It is the end of the world, since worlds depend on backgrounds and foregrounds. *World* is a fragile aesthetic effect around whose
corners we are beginning to see. True planetary awareness is the creeping
realization not that "We Are the World," but that we aren't.

Why? Because *world* and its cognates—*environment, nature*—are ironically
more objectified than the kinds of "object" I am talking about in this study.
World is more or less a container in which objectified things float or stand. It
doesn't matter very much whether the movie within the context of *world* is
an old-fashioned Aristotelian movie of substances decorated with accidents
or a more avant-garde Deleuzian one of flows and intensities. *World* as the
background of events is an objectification of a hyperobject: the biosphere,
climate, evolution, capitalism (yes, perhaps economic relations compose
hyperobjects). So when climate starts to rain on our head, we have no idea
what is happening. It is easy to practice denial in such a cognitive space: to
set up, for example, "debates" in which different "sides" on global warming

are presented. This taking of "sides" correlates all meaning and agency to the human realm, while in reality it isn't a question of sides, but of real entities and human reactions to them. Environmentalism seems to be talking about something that can't be seen or touched. So in turn environmentalism ups the ante and preaches the coming apocalypse. This constant attempt to shock and dismay inspires even more defiance on the opposite side of the "debate."

Both sides are fixated on *world,* just as both sides of the atheism debate are currently fixated on a *vorhanden* ("present at hand"), objectively present God. As irritating for New Atheists such as Richard Dawkins to hear that atheism is just another form of belief, it nevertheless is—or at any rate, it holds exactly the same *belief about belief* as the fundamentalists. Belief is a token, a mental object that you grip as hard as possible, like your wallet or car keys. In exactly the same way, it is annoying for environmentalists to talk about ecology without Nature. The argument is heard as nihilism or postmodernism. But really it is environmentalism that is nihilist and postmodernist, just as fundamentalism's belief about belief marks it as a form of ontotheological nihilism. The ultimate environmentalist argument would be to drop the concepts *Nature* and *world,* to cease identifying with them, to swear allegiance to coexistence with nonhumans without a world, without some nihilistic Noah's Ark.

In any weather conversation, one of you is going to mention global warming at some point. Or you both decide not to mention it but it looms over the conversation like a dark cloud, brooding off the edge of an ellipsis.[1]

This failure of the normal rhetorical routine, these remnants of shattered conversation lying around like broken hammers (they must take place everywhere), is a symptom of a much larger and deeper ontological shift in human awareness. And in turn, this is a symptom of a profound upgrade of our ontological tools. As anyone who has waited while the little rainbow circle goes around and around on a Mac, these upgrades are not necessarily pleasant. It is very much the job of philosophers and other humanities scholars to attune ourselves to the upgrading process and to help explain it.

What is the upgrading process? In a word, the notion that we are living "in" a world—one that we can call *Nature*—no longer applies in any meaningful sense, except as nostalgia or in the temporarily useful local language of pleas and petitions. We don't want a certain species to be farmed to extinction, so we use the language of Nature to convince a legislative body. We have a general feeling of ennui and malaise and create nostalgic visions of hobbit-like worlds to inhabit. These syndromes have been going on now since the Industrial Revolution began to take effect.

As a consequence of that revolution, however, something far bigger and more threatening is now looming on our horizon—looming so as to abolish our horizon, or any horizon. Global warming has performed a radical shift in the status of the weather. Why? Because *the world* as such—not just a specific idea of world but *world* in its entirety—has evaporated. Or rather, we are realizing that we never had it in the first place.

We could explain this in terms of the good old-fashioned Aristotelian view of substance and accidents. For Aristotle, a realist, there are *substances* that happen to have various qualities or *accidents* that are not intrinsic to their substantiality. In section Epsilon 2 of the *Metaphysics,* Aristotle outlines the differences between substances and accidents. What climate change has done is shift the weather from accidental to substantial. Aristotle writes, "Suppose, for instance, that in the season of the Cynosure [the Dog Days of summer] arctic cold were to prevail, this we would regard as an accident, whereas, if there were a sweltering heat wave, we would not. And this is because the latter, unlike the former, is always or for the most part the case."[2]

But these sorts of violent changes are exactly what global warming predicts. So every accident of the weather becomes a potential symptom of a substance, global warming. All of a sudden this wet stuff falling on my head is a mere feature of some much more sinister phenomenon that I can't see with my naked human eyes. I need terabytes of RAM to model it in real time (this has been available for about ten years).

There is an even spookier problem arising from Aristotle's arctic summer idea. If those arctic summers continue, and if we can model them as symptoms of global warming, then there *never was* a genuine, meaningful (for us humans) sweltering summer, just a long period of sweltering that seemed real because it kept on repeating for, say, two or three millennia. Global warming plays a very mean trick. It reveals that what we took to be a reliable world was actually just a habitual pattern—a collusion between forces such as sunshine and moisture and humans expecting such things at certain regular intervals and giving them names, such as Dog Days. We took weather to be real. But in an age of global warming we see it as an accident, a simulation of something darker, more withdrawn—climate. As Harman argues, *world* is always presence-at-hand—a mere caricature of some real object.[3]

Now let's think the evaporation of *world* from the point of view of *foreground* and *background*. A weather conversation provides a nice background to our daily affairs, nice to the extent that we don't pay too much attention to it. Precisely for it to be a background, it has to operate in our peripheral vision. Thus, the conversation about the weather with a stranger is a safe way to acknowledge our coexistence in social space. It's "phatic," according to Roman Jakobson's six-part model of communication; that is, it draws attention to the material medium in which the communication is occurring.[4] Likewise, the weather as such is a background phenomenon. It might loom distressingly into the foreground as a tornado or as a drought, but most often those are temporary affairs—there is a larger temporal backdrop against which they seem to occur as isolated incidents.

Now what happens when global warming enters the scene? The background ceases to be a background, because we have started to observe it. Strange weather patterns and carbon emissions caused scientists to start monitoring things that at first only appeared locally significant. That's the old definition of climate: there's the climate in Peru, the climate on Long Island, and so on. But climate in general, climate as the totality of derivatives of weather events—in much the same way as inertia is a derivative of velocity—is a beast newly recognized via the collaboration of weather, scientists, satellites, government agencies, and other entities. This beast includes the Sun, since it's infrared heat from the Sun that is trapped by the greenhouse effect of gases such as CO_2. So global warming is a colossal entity that includes entities that exist way beyond Earth's atmosphere, and yet it affects us intimately, right here and now. Global warming covers the entire surface of Earth, and 75 percent of it extends five hundred years into the future. Remember what life was like in the early 1500s?

Global warming is really here—even more spookily, it was already here, already influencing the supposedly real wet stuff falling on my head and the warm golden stuff burning my face at the beach. That wet stuff and that golden stuff, which we call weather, turns out to have been a false immediacy, an ontic pseudo-reality that can't stand up against the looming presence of an invisible yet far more real global climate. Weather, that handy backdrop for human lifeworlds, has ceased to exist, and along with it, the cozy concept of lifeworld itself. *Lifeworld* was just a story we were telling ourselves on the inside of a vast, massively distributed hyperobject called climate, a story about how different groups were partitioned according to different horizons—concepts now revealed as ontic prejudices smuggled into the realm of ontology. Global warming is a big problem, because along with melting glaciers it has melted our ideas of world and worlding. Thus, the tools that humanists have at their disposal for talking about the ecological emergency are now revealed, by global warming itself, to be as useless as the proverbial chocolate teapot. It is rather like the idea of using an antique (or better, antiqued) Christmas ornament as a weapon.

The spooky thing is, we discover global warming precisely when it's already here. It is like realizing that for some time you had been conducting your business in the expanding sphere of a slow-motion nuclear bomb. You have a few seconds for amazement as the fantasy that you inhabited a neat, seamless little world melts away. All those apocalyptic narratives of doom about the "end of the world" are, from this point of view, part of the problem, not part of the solution. By postponing doom into some hypothetical future, these narratives inoculate us against the very real object that has intruded into ecological, social, and psychic space. As we shall see, the hyperobject spells doom now, not at some future date. (*Doom* will assume a special technical meaning in this study in the "Hypocrisies" section.)

If there is no background—no neutral, peripheral stage set of weather, but rather a very visible, highly monitored, publicly debated climate—then there is no foreground. Foregrounds need backgrounds to exist. So the strange effect of dragging weather phenomena into the foreground as part of our awareness of global warming has been the gradual realization that there is no foreground! The idea that we are embedded in a phenomenological lifeworld, tucked up like little hobbits into the safety of our burrow, has been exposed as a fiction. The specialness we granted ourselves as unravelers of cosmic meaning, exemplified in the uniqueness of Heideggerian Dasein, falls apart since there is no meaningfulness possible in a world without a foreground–background distinction. Worlds need horizons and horizons need backgrounds, which need foregrounds. When we can see everywhere (when I can use Google Earth to see the fish in my mom's pond in her garden in London), the world—as a significant, bounded, horizoning entity—disappears. We have no world because the objects that functioned as invisible scenery have dissolved.[5]

World is an aesthetic effect based on a blurriness and aesthetic distance. This blurriness derives from ignorance concerning objects. Only in ignorance can objects act like blank screens for the projection of meaning. "Red sky at night, shepherd's delight" is a charming old saw that evokes days when shepherds lived in worlds bounded by horizons on which things such as red sunsets occurred. The Sun goes down, the Sun comes up—of course now we know it doesn't: Galileo and Copernicus tore holes in that notion of *world*. Likewise, as soon as humans know about climate, weather becomes a flimsy, superficial appearance that is a mere local representation of some much larger phenomenon that is strictly invisible. You can't see or smell climate. Given our brains' processing power, we can't even really think about it all that concretely. At the very least, *world* means significantly less than it used to—it doesn't mean "significant for humans" or even "significant for conscious entities."

A simple experiment demonstrates plainly that *world* is an aesthetic phenomenon. I call it *The Lord of the Rings* vs. the Ball Popper Test. For this experiment you will need a copy of *The Two Towers,* the second part of director Peter Jackson's *Lord of the Rings* trilogy.[6] You will also require a Playskool Busy Ball Popper, made by Hasbro. Now play the scene that I consider to be the absolute nadir of horror, when Frodo, captured by Faramir, is staggering around the bombed-out city Osgiliath when a Nazgûl (a ringwraith) attacks on a "fell beast," a terrifying winged dragon-like creature.

Switch on the Ball Popper. You will notice the inane tunes that the Popper plays instantly undermine the coherence of Peter Jackson's narrative world.

The idea of *world* depends on all kinds of mood lighting and mood music, aesthetic effects that by definition contain a kernel of sheer ridiculous meaninglessness. It's the job of serious Wagnerian worlding to erase the trace of this meaninglessness. Jackson's trilogy surely is Wagnerian, a total work of art *(Gesamtkunstwerk)* in which elves, dwarves, and men have their own languages, their own tools, their own architecture, done to fascist excess as if they were different sports teams. But it's easy to recover the trace of meaninglessness from this seamless world—absurdly easy, as the toy experiment proves. In effect, this stupid kids' toy "translated" the movie, clashing with it and altering it in its own limited and unique way.

Objections to wind farms and solar arrays are often based on arguments that they "spoil the view."[7] The aesthetics of Nature truly impedes ecology, and a good argument for why ecology must be without Nature. How come a wind turbine is less beautiful than an oil pipe? How come it "spoils the view" any more than pipes and roads? You could see turbines as environmental art. Wind chimes play in the wind; some environmental sculptures sway and rock in the breeze. Wind farms have a slightly frightening size and magnificence. One could easily read them as embodying the aesthetics of the sublime (rather than the beautiful). But it's an ethical sublime, one that says, "We humans choose not to use carbon"—a choice visible in gigantic turbines. Perhaps it's this very visibility of choice that makes wind farms disturbing: visible choice, rather than secret pipes, running under an apparently undisturbed "landscape" (a word for a painting, not actual trees and water). As a poster in the office of Mulder in the television series *The X-Files* says, "The Truth Is Out There." Ideology is not just in your head. It's in the shape of a Coke bottle. It's in the way some things appear "natural"—rolling hills and greenery—as if the Industrial Revolution had never occurred, and moreover, as if agriculture was Nature. The "landscape" look of agriculture is the original "greenwashing." Objectors to wind farms are not saying "Save the environment!" but "Leave our dreams undisturbed!" *World* is an aesthetic construct that depends on things like underground oil and gas pipes. A profound political act would be to choose another aesthetic construct, one that doesn't require smoothness and distance and coolness. *World* is by no means doing what it should to help ecological criticism. Indeed, the more data we have, the less it signifies a coherent world.

World is a function of a very long-lasting and complex set of social forms that we could roughly call the logistics of agriculture. New Zealand is an astonishing place where there are fifteen sheep for every human, a hyperbolic blowup of the English Lake District. It was deliberately manufactured that way. *World* is not just an idea in your head. It's in the way the fields roll toward a horizon, on top of which a red setting sun augurs peace and contentment. It's in the smooth, lawn-like texture of sheep-nibbled grass: "First the labourers are driven from the land, and then the sheep arrive."

Wind farms are an eyesore on this aestheticized landscape. Agriculture, in this view, is an ancient technological *world-picture,* to use Heidegger's terms: a form of framing that turns reality into so much stuff on tap *(Bestand).* [9] Agriculture is a major contributor to global warming, not just because of flatulent cows, but because of the enormous technical machinery that goes into creating the agricultural stage set, the *world.* Perhaps the solution to this is suggested by the kinds of "perverse" technologies developed by pot farmers: to create intensive growth in a small space. Just as the porn industry accelerated the development of the Internet, so the drug industry might be our ecological savior. Stranger things have happened. Preserving the agricultural world picture just as it is, however, has already become a costly disaster.

To return to an example close to New Zealand's heart, *The Lord of the Rings* presents an agricultural landscape that never explains itself. Sure, the Rangers such as Aragorn protect it. But how does it work? For whom and with whom is the growing and the harvesting and the selling done? Hobbiton is constructed to induce nostalgia for a suburban future that thinks itself as a Georgic idyll. To do so requires all kinds of lighting, rendering, and mood music—it also requires the threats of Mordor and orcs that make us care about bland suburbia. Just changing the Wagnerian music would destroy its delicate "balance."

Village Homes is a world-like real illusion that rests in the northwest area of Davis, California. Each street is named after a place or person in *The Lord of the Rings:* Evenstar Lane, Bombadil Lane. The streets are concentric yet nontopologically equivalent, so there is a real feeling of being lost in there. There are vineyards and pomegranate trees. There is a village-green-like space with an amphitheater built into the grass. There is a children's day care called Rivendell. It is all very beautiful; it's very well done. There is already a nostalgia for the present there, not simply for Tolkien, but for an ecological vision of the 1970s when Village Homes was designed. There is one slight problem: you have to have an awful lot of money to live there. And there is a rule that you have to work in the collective allotments. As a friend quipped, "One homeowner's association to rule them all."

There are many reasons why, even if *world* were a valid concept altogether, it shouldn't be used as the basis for ethics. Consider only this: witch-ducking stools constitute a world just as much as hammers. There was a wonderful world of witch ducking in the Middle Ages in which witches were "discovered" by drowning them, strapped to an apparatus that submerged them in the local stream: if the supposed witch didn't drown, she was a witch—and should thus be burned at the stake. Witch ducking stools constituted a world for their users in every meaningful sense. There is a world of Nazi regalia. Just because the Nazis had a world doesn't mean we should preserve it. So the argument that "It's good be—cause it constitutes a world" is flimsy at best. The reason not to interfere with the environment because it's interfering with someone's or something's world is nowhere near a good enough reason. It might even have pernicious consequences. *World* and *worlding* are a dangerously weak link in the series of late-Heideggerian concepts.[10] It is as if humans are losing both their world and their idea of *world* (including the idea that they ever had a world) at one and the same time, a disorienting fact. In this historical moment, working to transcend our notion of *world* is important. Like a mannerist painting that stretches the rules of classicism to a breaking point, global warming has stretched our *world* to breaking point. Human beings lack a world for a very good reason: because *no entity at all has a world*, or as Harman puts it, "There is no such thing as a 'horizon.'"[11] The "world" as the significant totality of what is the case is strictly unimaginable, and for a good reason: it doesn't exist.

What is left if we aren't the world? Intimacy. We have lost the world but gained a soul—the entities that coexist with us obtrude on our awareness with greater and greater urgency. Three cheers for the so-called end of the world, then, since this moment is the beginning of history, the end of the human dream that reality is significant for them alone. We now have the prospect of forging new alliances between humans and nonhumans alike, now that we have stepped out of the cocoon of *world*.

About six minutes into Pierre Boulez's piece *Répons,* the percussive instruments come in. They surround the smoother instruments (brass, strings), which are playing in a square in the center of the concert hall. The percussive instruments (piano, dulcimer, harp, and so on) are processed through various delays and filters. The sound of their entry is now evocative of speculative realism: the sound of a vaster world bursting into the human, or the reverse, the sound of a trapdoor opening in a plane, or the plane itself disappearing so we find ourselves in the wide blue sky. A terrifying, wonderful sound, the Kantian sublime of inner freedom giving way to a speculative sublime of disturbing intimacy. The sound of the end of the world but not an apocalypse, not a predictable conclusion. The sound of something beginning, the sound of discovering yourself inside of something.

Boulez himself probably thought *Répons* was about the sound of modern human technology, *Gesellschaft* (modern "society") impinging on *Gemeinschaft* (the "organic community"), and so forth. Or the idea of a dialogue between equal partners, a dialectical play between the organic and electronic. The piece is much more that that. It's the sound of real entities appearing to humans. But as I've been arguing, real nonhuman entities appear to humans at first as blips on their monitors. But they are not those blips. The sound of a higher-dimensional configuration space impinging on extreme Western music (total serialism). The sound of hyperobjects. The sound of a nonmusic. Listen to the very end: the sound echoes and reverberates, repeating glissandos; then, suddenly, it's over. No fade out. Robert Cahen captures it well in his deceptively simple film of *Répons,* visualizing the "human sounds" as a traditional orchestral ensemble juxtaposed with revolving and panning shots of trees, and the percussive sounds as humans mediated by *a luminous ocean.*[12] When the percussive instruments enter, the camera on the orchestra pans back to reveal them surrounding the other players, and we see the studio lighting rig, as if the structures that hold the fragile fiction of *world* together have evaporated. Just as most of Earth's surface is water, the sonic space is surrounded by the chilling, sparkling sounds of piano, harp, and glockenspiel.

Instead of trying constantly to tweak an illusion, thinking and art and political practice should simply relate directly to nonhumans. We will never "get it right" completely. But trying to come up with the best world is just inhibiting ecological progress. Art and architecture in the time of hyperobjects must (automatically) directly include hyperobjects, even when they try to ignore them. Consider the contemporary urge to maximize throughput: to get dirty air flowing with air conditioners. Air conditioning is now the benchmark of comfort; young Singaporeans are starting to sweat out of doors, habituated to the homogeneous thermal comfort of modern buildings.[13] Such architecture and design is predicated on the notion of "away." But there is no "away" after the end of the world. It would make more sense to design in a dark ecological way, admitting our coexistence with toxic substances we have created and exploited. Thus, in 2002 the architectural firm R&Sie designed *Dusty Relief,* an electrostatic building in Bangkok that would collect the dirt around it, rather than try to shuffle it somewhere else.[14] Eventually the building would be coated with a gigantic fur coat of dirt.[15]

Such new ideas are counterintuitive from the standpoint of regular post-1970s environmentalism. Process relationism has been the presiding deity of this thinking, insofar as it thinks flows are better than solids. But thinking this way on a planetary scale becomes absurd. Why is it better to stir the shit around inside the toilet bowl faster and faster rather than just leaving it there? Monitoring, regulating, and controlling flows: Is ecological ethics and politics just this? Regulating flows and sending them where you think they need to go is not relating to nonhumans. Regulation of flows is just a contemporary mode of window dressing of the substances of ontotheological nihilism, the becomings and processes with which Nietzsche wanted to undermine philosophy.

Contemporary architecture and design is thinking beyond models based on vectors and flow. When one considers Earth or the biosphere as a whole, pushing pollution "somewhere else" is only redistributing it, sweeping it under the carpet. Reproduced by permission. The common name for managing and regulating flows is *sustainability*. But what exactly is being sustained? "Sustainable capitalism" might be one of those contradictions in terms along the lines of "military intelligence."[16] Capital must keep on producing more of itself in order to continue to be itself. This strange paradox is fundamentally, structurally imbalanced. Consider the most basic process of capitalism: the turning of raw materials into products. Now for a capitalist, the raw materials are not strictly natural. They simply exist prior to whatever labor process the capitalist is going to exert on them. Surely here we see the problem. Whatever exists prior to the specific labor process is a lump that only achieves definition as valuable product once the labor has been exerted on it.

What capitalism makes is some kind of stuff called capital. The very definition of "raw materials" in economic theory is simply "the stuff that comes in through the factory door." Again, it doesn't matter what it is. It could be sharks or steel bolts. At either end of the process we have featureless chunks of stuff—one of those featureless chunks being human labor. The point is to convert the stuff that comes in to money. Industrial capitalism is philosophy incarnate in stocks, girders, and human sweat. What philosophy? If you want a "realism of the remainder," just look around you. "Realism of the remainder" means that yes, for sure, there is something real outside of our access to it—but we can only classify it as an inert resistance to our probing, a *gray goo,* to adapt a term suggested by thinking about nanotechnology—tiny machines eating everything until reality becomes said goo.

It's no wonder that industrial capitalism has turned the Earth into a dangerous desert. It doesn't really care what comes through the factory door, just as long as it generates more capital. Do we want to sustain a world based on a philosophy of gray goo? (Again, the term that some futurologists use to describe the nightmare of nano-scale robots mashing everything up into a colorless morass.) Nature is the featureless remainder at either end of the process of production. Either it's exploitable stuff, or value-added stuff. Whatever it is, it's basically featureless, abstract, gray. It has nothing to do with nematode worms and orangutans, organic chemicals in comets and rock strata. You can scour the earth, from a mountaintop to the Marianas Trench. You will never find Nature. It's an empty category looking for something to fill it.

Rather than only evaporating everything into a sublime ether (Marx via *Macbeth*: "All that is solid melts into air"), capitalism also requires and keeps firm long-term inertial structures such as families, as Fernand Braudel explored.[17] The Koch brothers and GE are two contemporary examples. One part of capital, itself a hyperobject, is its relentless revolutionizing of its mode of production. But the other part is tremendous inertia. And the tremendous inertia happens to be on the side of the modern. That is, the political ontology in which there is an "away." But there is no "away" in the time of hyperobjects.

Capitalism did away with feudal and prefeudal myths such as the divine hierarchy of classes of people. In so doing, however, it substituted a giant myth of its own: Nature. Nature is precisely the lump that exists prior to the capitalist labor process. Heidegger has the best term for it: *Bestand* (standing reserve). *Bestand* means "stuff," as in the ad from the 1990s, "Drink Pepsi: Get Stuff." There is an ontology implicit in capitalist production: materialism as defined by Aristotle. This specific form of materialism is not fascinated with material objects in all their manifold specificity. It's just stuff. This viewpoint is the basis of Aristotle's problem with materialism. Have you ever seen or handled matter? Have you ever held a piece of "stuff"? To be sure one has seen plenty of objects: Santa Claus in a department store, snowflakes, photographs of atoms. But have I ever seen matter or stuff as such? Aristotle says it's a bit like searching through a zoo to find the "animal" rather than the various species such as monkeys and mynah birds.[18] Marx says exactly the same thing regarding capital.[19] As Nature goes, so goes matter. The two most progressive physical theories of our age, ecology and quantum theory, need have nothing to do with it.

What is *Bestand*? *Bestand* is stockpiling. Row upon row of big box houses waiting to be inhabited. Terabyte after terabyte of memory waiting to be filled. Stockpiling is the art of the zeugma—the yoking of things you hear in phrases such as "wave upon wave" or "bumper to bumper." Stockpiling is the dominant mode of social existence. Giant parking lots empty of cars, huge tables in restaurants across which you can't hold hands, vast empty lawns. Nature is stockpiling. Range upon range of mountains, receding into the distance. Rocky Flats nuclear bomb trigger factory was sited precisely to evoke this mountainous stockpile. The eerie strangeness of this fact confronts us with the ways in which we still believe that Nature is "over there"—that it exists apart from technology, apart from history. Far from it. Nature is the stockpile of stockpiles.

What exactly are we sustaining when we talk about sustainability? An intrinsically out-of-control system that sucks in gray goo at one end and pushes out gray value at the other. It's Natural goo, Natural value. Result? Mountain ranges of inertia, piling higher every year, while humans boil away in the agony of uncertainty. Look at *Manufactured Landscapes*, the ocean of telephone dials, dials as far as the eye can see, somewhere in China.[20] Or consider the gigantic billowing waves of plastic cups created by Tara Donovan in *Untitled (Plastic Cups)* (2006). In massive piles, the cups reveal properties hidden from the view of a person who uses a single cup at a time, a viscous (in my terms) malleability. In Donovan's title, "cups" are in parentheses, the "untitled" outside parenthesis, as if to highlight the way the cups are "saying" something beyond their human use: something unspeakable for a human. The title of no-title places the work both inside and outside human social and philosophical space, like a garbage dump, an idea the gigantic pile surely evokes.

Figure 1. Tara Donovan, Untitled (Plastic Cups) (2006), plastic cups, dimensions variable. A billowing cloud of plastic made of mundane cups. Donovan plays with the disorienting way in which the human ability to calculate scale evokes strange entities that exist as much as a single plastic cup, but that occupy a dimension that is less available (or wholly unavailable) to mundane human perception. Photograph by Ellen Labenski. Copyright Tara Donovan, courtesy of Pace Gallery. Reproduced by permission.

Societies embody philosophies. What we have in modernity is considerably worse than just instrumentality. Here we must depart from Heidegger. What's worse is the location of essence in some *beyond,* away from any specific existence. To this extent, capitalism is itself Heideggerian!

Whether we call it scientism, deconstruction, relationism, or good old-fashioned Platonic forms, there is no essence in what exists. Either the beyond is itself nonexistent (as in deconstruction or nihilism), or it's some kind of real away from "here." The problem, then, is not essentialism but *this very notion of a beyond.* This beyond is what Tara Donovan's work destroys.

Tony Hayward was the CEO of BP at the time of the Deepwater Horizon oil pipe explosion, and his callousness made international headlines. Hayward said that the Gulf of Mexico was a huge body of water, and that the spill was tiny by comparison. Nature would absorb the industrial accident. I don't want to quibble about the difference in size between the Gulf and the spill, as if an even larger spill would somehow have gotten it into Hayward's thick head that it was bad news. I simply want to point out the metaphysics involved in Hayward's assertion, which we could call capitalist essentialism. The essence of reality is capital and Nature. Both exist in an ethereal beyond. Over here, where we live, is an oil spill. But don't worry. The beyond will take care of it.

Meanwhile, despite Nature, despite gray goo, real things writhe and smack into one another. Some leap out because industry malfunctions, or functions only too well. Oil bursts out of its ancient sinkhole and floods the Gulf of Mexico. Gamma rays shoot out of plutonium for twenty-four thousand years. Hurricanes congeal out of massive storm systems, fed by the heat from the burning of fossil fuels. The ocean of telephone dials mounts ever higher. Paradoxically, capitalism has unleashed myriad *objects* upon us, in their manifold horror and sparkling splendor. Two hundred years of idealism, two hundred years of seeing humans at the center of existence, and now the objects take revenge, terrifyingly huge, ancient, long-lived, threateningly minute, invading every cell in our body. When we flush the toilet, we imagine that the U-bend takes the waste away into some ontologically alien realm.[21] Ecology is now beginning to tell us about something very different: a flattened world without ontological U-bends. A world in which there is no "away." Marx was partly wrong, then, when in *The Communist Manifesto* he claimed that in capitalism all that is solid melts into air. He didn't see how a hypersolidity oozes back into the emptied-out space of capitalism. This oozing real can no longer be ignored, so that even when the spill is supposedly "gone and forgotten," there it is, mile upon mile of strands of oil just below the surface, square mile upon square mile of ooze floating at the bottom of the ocean.[22] It can't be gone and forgotten—even ABC News knows that now.

When I hear the word *sustainability* I reach for my sunscreen. The deep reason for why sustainability fails as a concept has to do with how we are not living in a *world*. It is thus time to question the very term *ecology*, since ecology is the thinking of home, and hence world (*oikos* plus *logos*). In a reality without a home, without world, what this study calls *objects* are what constitute reality. Objects are unique. Objects can't be reduced to smaller objects or dissolved upward into larger ones. Objects are withdrawn from one another and from themselves. Objects are Tardis-like, larger on the inside than they are on the outside. Objects are uncanny. Objects compose an untotalizable nonwhole set that defies holism and reductionism. There is thus no top object that gives all objects value and meaning, and no bottom object to which they can be reduced. If there is no top object and no bottom object, it means that we have a very strange situation in which there are more parts than there are wholes.[23] This makes holism of any kind totally impossible.

Even if you bracket off a vast amount of reality, you will find that *there is no top and bottom object in the small section you've demarcated*. Even if you select only a sector of reality to study somewhere in the middle, like they do in ecological science (the *mesocosm*), you will also find no top or bottom object, even as it pertains to that sector alone. It's like a magnet. If you cut it, the two halves still have a north and a south pole. There is no such thing as "half" a magnet versus a "whole" one.

Why is holism such a bad idea? Surely there could be other possible holisms that adopt some version of both–and thinking so that neither the parts nor the whole—whatever the whole might be—are greater. Perhaps the parts are not necessarily lesser than the whole but exist in some both–and synergistic fashion; you could have—simultaneously—"withdrawn" objects and something else (just to satisfy our modern need for things that aren't static, let's say an open-ended, possibly always expanding, something else).

First, we must walk through some semirelated points about this line of questioning. It sounds like good value to have "both–and" rather than "either–or," to our somewhat consumerist minds ("buy one get one free"). But I'm afraid this is a case of either–or: holism or not. The parts are *not* replaceable components of the whole. The more we open up the Russian doll of an object, the more objects we find inside. Far more than the first object in the series, because *all the relations between the objects and within them also count as objects*. It's what Lacanians call a not-all set. Objects in this sense are fundamentally not subject to phallogocentric rule. (Commercial break: If you're having trouble with *object* at this point, why not try another term, such as *entity*?) What we encounter in OOO, which I have been expounding in these last couple of pages, is a Badiou-like set theory in which any number of affiliations between objects can be drawn. The contents of these sorts of sets are bigger than the container.

Sometimes children's books explore deep ontological issues. The title of *A House Is a House for Me* couldn't be better for a book about ecology (see my observation above about *oikos* and *logos*). The text is a wonderfully jumbly plethora of objects: Cartons are houses for crackers. Castles are houses for kings. The more that I think about houses, the more things are houses for things.[24] Home, *oikos,* is unstable. Who knows where it stops and starts? The poem presents us with an increasingly dizzying array of objects. They can act as homes for other objects. And of course, in turn, these homes can find themselves on the inside of other "homes."

Home is purely "sensual": it has to do with how an object finds itself inevitably on the inside of some other object. The instability of *oikos,* and thus of ecology itself, has to do with this feature of objects. A *house* is the way an object experiences the entity in whose interior it finds itself. So then these sorts of things are also houses: A mirror's a house for reflections . . . A throat is a house for a hum . . . A book is a house for a story. A rose is a house for a smell. My head is a house for a secret, a secret I never will tell. A flower's at home in a garden. A donkey's at home in a stall. Each creature that's known has a house of its own and the earth is a house for us all.[25]

The time of hyperobjects is the time during which we discover ourselves on the inside of some big objects (bigger than us, that is): Earth, global warming, evolution. Again, that's what the *eco* in *ecology* originally means: *oikos,* home. The last two lines of *A House Is a House for Me* make this very clear.

To display the poem's effortless brio, a lot of silly, fun "houses" are presented in the penultimate section as we hurry toward the conclusion, which then sets the record straight by talking about a "real" house, the Earth. But this is not the case. OOO doesn't claim that any object is "more real" than any other. But it does discount some objects, which it calls *sensual objects.* What is a sensual object? *A sensual object is an appearance-for another object.* The table-for my pencil is a sensual object. The table-for my eyes is a sensual object. The table-for my dinner is a sensual object. Sensual objects are wonderfully, disturbingly entangled in one another. *This is where causality happens,* not in some mechanical basement. This is where the magical illusion of appearance happens. *A mirror's a house for reflections.* Yes, *the mesh* (the interrelatedness of everything) is a sensual object! Strange strangers are the real objects! Some very important entities that environmentalism thinks of as real, such as Nature, are also sensual objects. They appear "as" what they are *for an experiencer or user or apprehender.* They are manifestations of what Harman calls the *asstructure.*[26] They are as-structured even though they appear to be some deep background to (human) events.

This confusion of sensual and real, in the terms of *A House Is House for Me,* is like thinking that bread *really is* a house for jam, and jam alone. Rather than simply an idea that occurs to me, and perhaps to the jam, when it finds itself slathered in there. Marmalade wants in on the bread? Too bad, marmalade is an artificial, unnatural parasite! Peanut butter? Illegal alien! Only jam is "natural," such that bread is only made-for jam. See the problem with

Nature? In OOO-ese, *reification* is precisely *the reduction of a real object to its sensual appearance-for another object.* Reification is the reduction of one entity to another's *fantasy* about it.

Nature is a reification in this sense. That's why we need ecology without Nature. Maybe if we turn Nature into something more fluid, it would work. Emergence is also a sensual object. And thus it's in danger of doing the work of reifying—strangely enough, given its reputation as an unreified, flowy thing, despite its popularity as a replacement for terms such as *nature*. Emergence is always emergence-for. Yet there is a deeper way to think emergence. *Physis, emergence, sway, the way a flower unfurls, seeming, upsurge of Being,* are some of the terms Heidegger uses to characterize what he considers to be the primary notion of the ancient Greek philosophers. There is an appearing-to, an emerging-for, going on. Being is not separated from seeming, at the most fundamental stage of Heidegger's account. And so there is no reason why a poem can't be construed as a physical object in as rich a sense as you like. It's only counterintuitive if you think that entities come with two floors: basement mechanics and a pretty living room on top. But for OOO, Heidegger's terms for being are simply elaborations on the as-structure. Whether you call it emergence or appearance, what we are talking about is a sensual object.

Thinking on a planetary scale means waking up inside an object, or rather a series of "objects wrapped in objects": Earth, the biosphere, climate, global warming.[27] Ecological being-with does not mean dusting some corner of an object so one doesn't feel too dirty. Ecological being-with has to do with acknowledging a radical uniqueness and withdrawal of things, not some vague sludge of *apeiron* (using Anaximander's term for "the limitless"). A circle, not an endless line, is a better emblem for the constraint, yet openness, of things.[28] Indeed, the vague sludge is precisely the problem of pollution. Process relationism is simply the last philosophical reflex of the modernity that creates the sludge. We need a philosophy of sparkling unicities; quantized units that are irreducible to their parts or to some larger whole; sharp, specific units that are not dependent on an observer to make them real.

These are considerations concerning the normative value of different ontologies. But there is a deeper reason why hyperobjects are best seen not as processes, but as real entities in their own right. Seen from a suitably high dimension, a process just is a static object. I would appear like a strange worm with a cradle at one end and a grave at the other, in the eyes of a four-dimensional being. This is not to see things *sub specie aeternitatis,* but as I argued previously, *sub specie majoris:* from a slightly higher-dimensional perspective. Processes are sophisticated from a lower-dimensional viewpoint. If we truly want to transcend anthropocentrism, this might not be the way to go. To think some things as processes is ironically to reify them as much as the enemy of the process philosopher supposedly sees things as static lumps. As static lumps go, Lorenz Attractors are pretty cool. Processes are equally reifications of real entities. A process is a sensual translation, a parody of a higherdimensional object by a lower-dimensional being. A hyperobject is like a city—indeed a city such as London could provide a good example of a hyperobject. Cities and hyperobjects are full of strange streets, abandoned entrances, cul-de-sacs, and hidden interstitial regions.

The Nuclear Guardianship movement advocates an approach to nuclear materials that is strikingly similar to the way in which the electrostatic building simply accumulates dirt without shunting it under the rug.[29] There is no *away* to which we can meaningfully sweep the radioactive dust. Nowhere is far enough or long-lasting enough. What must happen instead is that we must care consciously for nuclear materials, which means keeping them above ground in monitored retrievable storage until they are no longer radioactive. Remember that the half-life of plutonium-239 is 24,100 years. That's almost as long into the future as the Chauvet Cave paintings are in our past. The future of plutonium exerts a causal influence on the present, casting its shadow backward through time. All kinds of options are no longer thinkable without a deliberate concealment of the reality of radioactive objects. Far, far more effort must be put into monitored retrievable storage than Thomas Sebeok's disturbing idea of an "atomic priesthood" that enforces ignorance about the hyperobject in question.[30] The documentary *Into Eternity* explores the immense challenge that the now immense heap of nuclear materials on Earth pose to thinking and to democracy.[31] The film is narrated for a far future addressee, displacing the spurious now, which we habitually think as a point or a small, rigid bubble.

Guardianship, care—to *curate* is to care for. We are the curators of a gigantic museum of non-art in which we have found ourselves, a spontaneous museum of hyperobjects. The very nature of democracy and society—Whom does it contain? Only humans? Whom, if any, can it exclude?—is thrown into question. The atomic priesthood would prevent others from knowing the truth.[32] The attempt to care for hyperobjects and for their distant future guardians will strikingly change how humans think about themselves and their relationships with nonhumans. This change will be a symptom of a gradually emerging ecological theory and practice that includes social policy, ethics, spirituality, and art, as well as science. Humans become, in Heidegger's words, the guardians of futurality, "the stillness of the passing of the last god."[33] Nuclear Guardianship has suggested encasing plutonium in gold, that precious object of global reverence and lust, rather than sweeping it away out of view. Encased in gold, which has the advantage of absorbing gamma rays, plutonium could become an object of contemplation. Set free from use, plutonium becomes a member of a democracy expanded beyond the human. Nature as such is a byproduct of automation. By embracing the hyperobjects that loom into our social space, and dropping Nature, *world,* and so on, we have a chance to create more democratic modes of coexistence between humans and with nonhumans. But these modes are not discernible within traditional Western parameters, since future generations—and further futures than that, are now included on "this" side of any ethical or political decision.[34]

Nuclear Guardianship sees nuclear materials as a unit: a hyperobject. This vision summons into human fields of thinking and action something that is already there. The summoning is to nuclear materials to join humans in social space, rather than remain on the outside. Or better, it's an acknowledgment by humans that nuclear materials are already occupying social space. It's an intrinsically scary thought. But wishing not to think it is just postponing the inevitable. To wish this thought away is tantamount to the cleanup operations that simply sweep the contaminated dust, garbage, and equipment away to some less politically powerful constituency. As a member of society, nuclear materials are a unit, a quantum that is not reducible to its parts or reducible upward into some greater whole. Nuclear materials constitute a unicity: *finitude* means just this. Nuclear materials may present us with a very large finitude, but not an infinitude. They simply explode what we mean by finitude. They are not objective lumps limited in time and space, but unique beings.[35] They have everything that Heidegger argues is unique to Dasein.

Hyperobjects are *futural,* as the section "Interobjectivity" demonstrated. They scoop out the objectified now of the present moment into a shifting uncertainty. Hyperobjects loom into human time like the lengthening shadow of a tree across the garden lawn in the bright sunshine of an ending afternoon. The end of the world is not a sudden punctuation point, but rather it is a matter of deep time. Twenty-four thousand years into the future, no one will be meaningfully related to me. Yet everything will be influenced by the tiniest decisions I make right now.[36] Inside the hyperobject nuclear radiation, I am like a prisoner, and a future person is like another prisoner. We are kept strictly apart, yet I guess his existence. There is a rumor going around the prison. If I make a deal with the police and pin the blame for my crime on the other prisoner, and he says nothing, I can go free and he receives a longer sentence. However, if I say nothing and he says nothing together, we both get a minor sentence. Yet if we both betray the other, we receive an even longer sentence. I can never be sure what the other will do. It would be optimal if I emphasize my self-interest above all other considerations. Yet it would be best if I act with a regard to the wellbeing of the other prisoner.

This is the Prisoner's Dilemma. In 1984 Derek Parfit published the groundbreaking *Reasons and Persons,* a book that exploded long-held prejudices about utility and ethics from within utilitarianism itself. Parfit showed that no self-interest ethical theory, no matter how modified, can succeed against such dilemmas.[37] Specifically Parfit has in mind hyperobjects, things such as pollution and nuclear radiation that will be around long after anyone meaningfully related to me exists. Since in turn my every smallest action affects the future at such a range, it is as if with every action I am making a move in a massive highly iterated Prisoner's Dilemma game. We might as well rename it Jonah's Dilemma or the Dilemma of the Interior of a Hyperobject. Default capitalist economics is rational choice theory, which is deeply a self-interest theory. Yet the Prisoner's Dilemma indicates we're profoundly social beings. Even self-interest accounts for the other somehow.

Parfit subjects an astonishing array of self-interest theories (variously modified to include relatives, friends, neighbors, descendants, and so on) to numerous tests based on the Prisoner's Dilemma. The Prisoner's Dilemma encourages one to think about how change begins: one thinks of the other, one brings the other into decisions that are supposedly about one's self-interest. To this extent the Prisoner's Dilemma is formally collectivist even though it lacks a positive collectivist or socialist content. The kinds of compromise necessitated by the Prisoner's Dilemma may strike ideological purists as weak. It is precisely this weakness that makes the so-called compromises workable and just. Imagine a future self with interests so different from one's own that to some extent she or he constitutes a different self: not your reincarnation or someone else—you yourself. This person in the future is like the prisoner being interrogated in the other room. The future self is thus unimaginably distant in one sense, and yet hyperobjects have brought her into the adjoining prison cell. She is strange yet intimate. The best course of action is to act with regard to her. This radical letting go of what constitutes a self has become necessary because of hyperobjects. The "weakness" of this ethical position is determined by the radical withdrawal of the future being: I can never fully experience, explain, or otherwise account for her, him, or it. The end of the world is a time of weakness.

The ethics that can handle hyperobjects is directed toward the unknown and unknowable future, the future that Jacques Derrida calls *l'avenir*.[38] Not the future we can predict and manage, but an unknowable future, a genuinely *future future*. In the present moment, we must develop an ethics that addresses what Derrida calls *l'arrivant,* the absolutely unexpected and unexpectable arrival, or what I call the *strange stranger,* the stranger whose strangeness is forever strange—it cannot be tamed or rationalized away. This stranger is not so unfamiliar: uncanny *familiarity* is one of the strange stranger's traits. Only consider anyone who has a long-term partner: the person they wake up with every day is the strangest person they know. The future future and the strange stranger are the weird and unpredictable entities that honest ecological thinking compels us to think about. When we can see that far into the future and that far around Earth, a curious blindness afflicts us, a blindness far more mysterious than simple lack of sight, since we can precisely see so much more than ever. This blindness is a symptom of an already existing intimacy with all lifeforms, knowledge of which is now thrust on us whether we like it or not.

Parfit's assault on utilitarian self-interest takes us to the point at which we realize that we are not separate from our world. Humans must learn to care for fatal substances that will outlast them and their descendants beyond any meaningful limit of self-interest. What we need is an ethics of the other, an ethics based on the proximity of the stranger. The decision in the 1990s, rapidly overturned, to squirrel plutonium away into knives and forks and other domestic objects appears monstrous, and so would any attempt to "work" it into something convenient. Hyperobjects insist that we care for them in the open. "Out of sight, out of mind" is strictly untenable. There is no "away" to throw plutonium in. We are stuck with it, in the same way as we are stuck with our biological bodies. Plutonium finds itself in the position of the "neighbor" in Abrahamic religions—that awkward condition of being alien and intimate at the very same time.

The enormity of very large finitude hollows out my decisions from the inside. Now every time I so much as change a confounded light bulb, I have to think about global warming. It is the end of the world, because I can see past the lip of the horizon of human worlding. Global warming reaches into "my world" and forces me to use LEDs instead of bulbs with filaments. This aspect of the Heideggerian legacy begins to teeter under the weight of the hyperobject. The normative defense of worlds looks wrongheaded.[39] The ethical and political choices become much clearer and less divisive if we begin to think of pollution and global warming and radiation as effects of hyperobjects rather than as flows or processes that can be managed. These flows are often eventually shunted into some less powerful group's backyard. The Native American tribe must deal with the radioactive waste. The African American family must deal with the toxic chemical runoff. The Nigerian village must deal with the oil slick. Rob Nixon calls this the *slow violence* of ecological oppression.[40] It is helpful to think of global warming as something like an ultra slow motion nuclear bomb. The incremental effects are almost invisible, until an island disappears underwater. Poor people—who include most of us on Earth at this point—perceive the ecological emergency not as degrading an aesthetic picture such as *world* but as an accumulation of violence that nibbles at them directly.

Without a world, there are simply a number of unique beings (farmers, dogs, irises, pencils, LEDs, and so on) to whom I owe an obligation through the simple fact that existence is coexistence. I don't have to run through my worlding checklist to ensure that the nonhuman in question counts as something I could care for. "If you answered mostly (A), then you have a world. If you answered mostly (B), then you are poor in world (German, *weltarm*). If you answered mostly (C), then you have no world whatsoever." What remains without a world is intimacy. Levinas touches on it in his ethics of alterity, although he is incorrect to make this otherness as vague as the "rustling" of blank existence, the "there is" *(il y a)*.[41] The other is fully here, before I am, as Levinas argues. But the other has paws and sharp surfaces, the other is decorated with leaves, the other shines with starlight. Kafka writes:

> At first glance it looks like a flat star-shaped spool for thread, and indeed it does seem to have thread wound upon it; to be sure, they are only old, broken-off bits of thread, knotted and tangled together, of the most varied sorts and colors. But it is not only a spool, for a small wooden crossbar sticks out of the middle of the star, and another small rod is joined to that at a right angle. By means of this latter rod on one side and one of the points of the star on the other, the whole thing can stand upright as if on two legs.[42]

"The idea that he is likely to survive me I find almost painful."[43] Kafka's Odradek resembles the hyperobject in this respect. Indeed we have let him into our home somehow, like mercury and microwaves, like the ultraviolet rays of the Sun. Odradek is what confronts us at the end of the world, not with a shout but with a breathless voice "like the rustling of fallen leaves."[44] Things appear in their disturbing *weakness* and *lameness,* technical terms describing the human attunement to hyperobjects that I have begun to elucidate.

Without a world, there is no Nature. Without a world, there is no life. What exists outside the charmed circles of Nature and life is a *charnel ground,* a place of life and death, of death-in-life and life-in-death, an undead place of zombies, viroids, junk DNA, ghosts, silicates, cyanide, radiation, demonic forces, and pollution. My resistance to ecological awareness is a resistance to the charnel ground. It is the calling of the shaman to enter the charnel ground and to try to stay there, to pitch a tent there and live there, for as long as possible. Since there are no charnel grounds to speak of in the West, the best analogy, used by some Tibetan Buddhists (from whom the image derives), is the emergency room of a busy hospital. People are dying everywhere. There is blood and noise, equipment rushing around, screams. When the charm of *world* is dispelled, we find ourselves in the emergency room of ecological coexistence.

In the charnel ground, worlds can never take root. Charnel grounds are too vivid for that. Any soft focusing begins to look like violence. Haunting a charnel ground is a much better analogy for ecological coexistence than inhabiting a world. There is something immensely soothing about charnel grounds. It is what is soothing about Buddhism's First Noble Truth, the truth of suffering. Traditionally, Buddhism recognizes three types of suffering. There is the pain of pain, as when you hit your thumb with a hammer, and then you close your whole hand in the door as you rush into the car to get to the doctor's because of your thumb. Then there is the pain of alteration, in which you experience first pleasure, then pain when pleasure evaporates. Then there is "all-pervasive pain," which Chögyam Trungpa beautifully describes as a "fundamental creepy quality" akin to Heidegger's description of Angst.[45] It is this quality that comes close to the notion of *world*. All-pervasive pain has to do with the fixation and confusion that constitute the Six Realms of Existence (traditionally, animals, humans, gods, jealous gods, hungry ghosts, and hell). In paintings of the Wheel of Life, the Six Realms are held in the jaws of Yama, the Lord of Death.

It is this outermost perspective of the jaws of death that provides an entry point into the charnel ground. To a Buddhist, ecophenomenological arguments that base ethics on our embeddedness in a lifeworld begin to look like a perverse aestheticization, celebrations of confusion and suffering for confusion's and suffering's sake. It doesn't really matter what is on the TV (murder, addiction, fear, lust). Each realm of existence is just a TV show taking up "space" in the wider space of the charnel ground of reality, "the desert of the real."[46] Trebbe Johnson and others have established the practice of Global Earth Exchanges, actions of finding, then giving something beautiful in a "wounded place," such as a toxic dump or a nuclear power facility.[47] Or consider Buddhist practitioners of tonglen: "sending and taking," a meditation practice in which one breathes out compassion for the other, while breathing in her or his suffering. Tonglen is now used in the context of polluted places. Consider Chöd, the esoteric ritual of visualizing cutting oneself up as a feast for the demons, another practice that has been taken on with reference to ecological catastrophes. Or consider the activities of Zen priests at the Rocky Flats nuclear bomb trigger factory, such as walking meditation. Our actions build up a karmic pattern that looks from a reified distance like a realm such as hell or heaven. But beyond the violence that we do, it's the distance that reifies the pattern into a *world picture* that needs to be shattered. Whether it's Hobbiton, or the jungles of *Avatar*, or the National Parks and conservation areas over yonder on the hither side of the screen

(though possibly behind the windshield of an SUV), or the fields and irrigation channels on the hither side of the wilderness—it's all a world picture.
I'm not saying we need to uproot the trees—I'm saying that we need to
smash the aestheticization: in case of ecological emergency, break glass.

Our increasing knowledge of global warming ends all kinds of ideas,
but it creates other ones. The essence of these new ideas is the notion of
coexistence—that is after all what ecology profoundly means. We coexist with human lifeforms, nonhuman lifeforms, and non-lifeforms, on the
insides of a series of gigantic entities with whom we also coexist: the ecosystem, biosphere, climate, planet, solar system. A multiple series of nested
Russian dolls. Whales within whales within whales.

Consider the hypothetical planet Tyche, far out in the Oort Cloud beyond
Pluto. We can't see it directly but we can detect evidence of its possible existence. Planets are hyperobjects in most senses. They have Gaussian geometry and measurable space-time distortion because they are so massive.
They affect everything that exists on and in them. They're "everywhere and
nowhere" up close *(viscosity)*. (Point to Earth right now—you have a number
of options of where to point.) They are really old and really huge compared with humans. And there's something disturbing about the existence of
a planet that far away, perhaps not even of "our" solar system originally, yet
close enough to be uncanny (a very large finitude). And it's unseen except
for its hypothetical influence on objects such as comets: "The awful shadow
of some unseen power," in Shelley's words. Tyche is a good name. It means
"contingency" in Greek, so it's the speculative realism planet par excellence.
("Luck" and "chance" are rather tame alternative translations. Tyche is what
happens to you in a tragedy if your name is Oedipus.) And for now, what
could be more obviously withdrawn?

The historic moment at which hyperobjects become visible by humans
has arrived. This visibility changes everything. Humans enter a new age of
sincerity, which contains an intrinsic irony that is beyond the aestheticized,
slightly plastic irony of the postmodern age. What do I mean?

This is a momentous era, at which we achieve what has sometimes been
called ecological awareness. Ecological awareness is a detailed and increasing sense, in science and outside of it, of the innumerable interrelationships
among lifeforms and between life and non-life. Now this awareness has some
very strange properties. First of all, the awareness ends the idea that we are

living in an environment! This is so bizarre that we should dwell on it a little. What it means is that the more we know about the interconnection, the more it becomes impossible to posit some entity existing beyond or behind the interrelated beings. When we look for the environment, what we find are discrete lifeforms, non-life, and their relationships. But no matter how hard we look, we won't find a container in which they all fit; in particular we won't find an umbrella that unifies them, such as world, environment, ecosystem, or even, astonishingly, Earth.

What we discover instead is an open-ended mesh that consists of grass, iron ore, Popsicles, sunlight, the galaxy Sagittarius, and mushroom spores. Earth exists, no doubt, but not as some special enormous bowl that contains all the "ecological" objects. Earth is one object coexisting with mice, sugar, elephants, and Turin. Of course there are many scenarios in which if Earth ceased to exist, Turin and mice would be in trouble. But if the mice were shot into space aboard a friendly extraterrestrial freighter, Earth wouldn't be the cause of their death. Even Turin might be rebuilt, brick by brick, on some other world.

Suddenly we discover the second astonishing thing. Mice are surely mice no matter what we call them. But mice remain mice as long as they survive to pass on their genome—it's what neo-Darwinism calls *satisficing*. Satisficing is a performative standard for existing. And there is no mouse-flavored DNA. There isn't even any DNA-flavored DNA—it's a palimpsest of mutations, viral code insertions, and so on. There isn't even any life-flavored life. DNA requires ribosomes and ribosomes require DNA, so to break the vicious cycle, there must have been an RNA world of RNA attached to a nonorganic replicator, such as a silicate crystal. So there is a mouse—this is not a nominalist nor is it an idealist argument. But the mouse is a non-mouse, or what I call a *strange stranger*.[48] Even more weirdly: this is why the mouse is real. The fact that wherever we look, we can't find a mouse, is the very reason why she exists! Now we can say this about everything in the universe. But one of the most obvious things we can say this about is a hyperobject. Hyperobjects are so huge and so long-lasting, compared with humans, that they obviously seem both vivid and slightly unreal, *for exactly the same reasons.*

Hyperobjects such as global warming and nuclear radiation surround us, not some abstract entity such as Nature or environment or *world.* Our reality has become more real, in the sense of more vivid and intense, and yet it has also become less knowable as some one-sided, facile thing—again, for exactly the same reasons. In Berkeley, California, in early 2011, radiation levels in water spiked 181 times higher than normal because of the Sendai reactor meltdowns. We know this. We know we are bathed in alpha, beta, and gamma rays emanating from the dust particles that now span the globe. These particles coexist with us. They are not part of some enormous bowl called Nature; they are beings like us, strange strangers.

Should we stop drinking water? Should we stop drinking cow's milk because cows eat grass, which drinks rainwater? The more we know, the harder it is to make a one-sided decision about anything. As we enter the time of hyperobjects, Nature disappears and all the modern certainties that seemed to accompany it. What remains is a vastly more complex situation that is uncanny and intimate at the same time.

There is no exit from this situation. Thus the time of hyperobjects is a time of sincerity: a time in which it is impossible to achieve a final distance toward the world. But for this very reason, it is also a time of irony. We realize that nonhuman entities exist that are incomparably more vast and powerful than we are, and that our reality is caught in them. What things are and how they seem, and how we know them, is full of gaps, yet vividly real. Real entities contain time and space, exhibiting nonlocal effects and other interobjective phenomena, writing us into their histories. Astonishingly, then, the mesh of interconnection is secondary to the strange stranger. The mesh is an emergent property of the things that coexist, and not the other way around. For the modernist mind, accustomed to systems and struc-tures, this is an astounding, shocking discovery. The more maps we make, the more real things tear through them. Nonhuman entities emerge through our mapping, then they destroy them.

Coexistence is in our face: it *is* our face. We are made of nonhuman and nonsentient and nonliving entities. It's not a cozy situation: it's a spooky, uncanny situation. We find ourselves in what robotics and CGI designers call *the uncanny valley* (Figure 2). It's a commonly known phenomenon in CGI design that if you build figures that look too much like humans, you are at risk of crossing a threshold and falling into the uncanny valley. In the uncanny valley, beings are strangely familiar and familiarly strange. The valley seems to explain racism quite well, because the dehumanization suf-fered by victims of racism makes them more uncanny to the racist than, say, a dog or a faceless robot. Hitler was very fond of his dog Blondi and yet dehumanized Jews and others. That's the trouble with some kinds of environmentalist language: they skip blithely over the uncanny valley to shake hands with beings on the other side. But, as I'm going to argue, there is only another side if you are holding on to some fictional idea of humanness, an idea that ecological awareness actually refutes. The uncanny valley, in other words, is only a valley if you already have some quite racist assumptions about lifeforms.

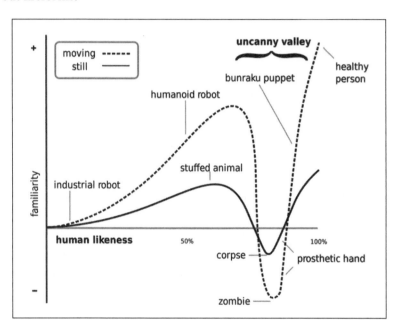

Figure 2. Masahiro Mori's diagram of the uncanny valley. Intimacy implies the grotesque. Since ecological awareness consists in a greater intimacy with a greater number of beings than modernity is capable of thinking, humans must pass through the uncanny valley as they begin to engage these beings. For reasons given in the book, this valley might be infinite in extent.

With ecological awareness there is no "healthy person" on the other side of the valley. Everything in your world starts to slip into the uncanny valley, whose sides are infinite and slick. It's more like an uncanny charnel ground, an ER full of living and dying and dead and newly born people, some of whom are humans, some of whom aren't, some of whom are living, some of whom aren't. Everything in your world starts to slip into this charnel ground situation, *including your world.*

Isn't it strange that we can admire comets, black holes, and suns—entities that would destroy us if they came within a few miles of us—and we can't get a handle on global warming? Isn't global climate now in the uncanny valley? Doesn't this have something to do with art? Because when you look at the stars and imagine life on other planets, you are looking through the spherical glass screen of the atmosphere at objects that appear to be behind that glass screen—for all the developments since Ptolemy, in other words, you still imagine that we exist on the inside of some pristine glass sphere. The experience of cosmic wonder is an aesthetic experience, a three-dimensional surround version of looking at a picturesque painting in an art gallery. So Jane Taylor's Romantic period poem "The Star" is about seeing stars through the atmosphere, in which they seem to "Twinkle, twinkle."

Two and a half thousand people showed up at the University of Arizona in Tucson for a series of talks on cosmology.[49] Evidently there is a thirst for thinking about the universe as a whole. Why is the same fascination not there for global warming? It's because of the oppressive claustrophobic horror of actually being inside it. You can spectate "the universe" as an ersatz aesthetic object: you have the distance provided by the biosphere itself, which acts as a spherical cinema screen. Habit tells us that what's displayed on that screen (like projections in a planetarium) is infinite, distant—the whole Kantian sublime. But inside the belly of the whale that is global warming, it's oppressive and hot and there's no "away" anymore. And it's profoundly regressing: a toxic intrauterine experience, on top of which we must assume responsibility for it. And what neonatal or prenatal infant should be responsible for her mother's existence? Global warming is in the uncanny valley, as far as hyperobjects go. Maybe a black hole, despite its terrifying horror, is so far away and so wondrous and so fatal (we would simply cease to exist anywhere near it) that we marvel at it, rather than try to avoid thinking about it or feel grief about it. The much smaller, much more immediately dangerous hole that we're in (inside the hyperobject global warming) is profoundly disturbing, especially because we created it.

Now the trouble with global warming is that it's right here. It's not behind a glass screen. It *is* that glass screen, but it's as if the glass screen starts to extrude itself toward you in a highly uncanny, scary way that violates the normal aesthetic propriety, which we know about from philosophers such as Kant—the propriety in which there should be a Goldilocks distance between you and the art object, not too close, not too far away. Global warming plays a very mean trick. It comes very, very close, crashing onto our beaches and forcing us to have cabinet meetings underwater to draw attention to our plight, and yet withdrawing from our grasp in the very same gesture, so that we can only represent it by using computers with tremendous processing speed.[50] The whale that Jonah is inside is a higher-dimensional being than ourselves, like two-dimensional stick people relative to a three-dimensional apple. We see that we are *weak,* in the precise sense that our discourse and maps and plans regarding things are not those things. There is an irreducible gap.

Spookily, the picture frame starts to melt and extrude itself toward us, it starts to burn our clothing. This is not what we paid twelve bucks to see when we entered the art gallery. Human art, in the face of this melting glass screen, is in no sense public relations. It has to actually *be* a science, part of science, part of cognitively mapping this thing. Art has to be part of the glass itself because everything inside the biosphere is touched by global warming.

Notes

1. Timothy Morton, *The Ecological Thought* (Cambridge, MA: Harvard University Press, 2010), 28, 54.

2. Aristotle, *Metaphysics,* trans. Hugh Lawson-Tancred (London: Penguin, 1999), 158–59.

3. Graham Harman, *Tool-Being: Heidegger and the Metaphysics of Objects* (Chicago: Open Court, 2002), 127.

4. Roman Jakobson, "Closing Statement: Linguistics and Poetics," in *Style in Language,* ed. Thomas A. Sebeok (Cambridge, MA: MIT Press, 1960), 350–77.

5. Harman, *Tool-Being,* 21–22.

6. *The Two Towers,* directed by Peter Jackson (New Line Cinema, 2002).

7. Anon, "Residents Upset about Park Proposal," *Lakewood Sentinel,* July 31, 2008, 1; "Solar Foes Focus in the Dark," *Lakewood Sentinel,* August 7, 2008, 4.

8. Karl Marx, *Capital, trans. Ben Fowkes, 3 vols. (Harmondsworth: Penguin, 1990),* 1:556.

9. Martin Heidegger, "The Question Concerning Technology," in *The Question Concerning Technology and Other Essays,* trans. William Lovitt (New York: Harper & Row, 1977), 17.

10. See, for instance, Heidegger, "Origin," 15–86.

11. Harman, *Tool-Being,* 155.

12. Pierre Boulez, *Répons* (Deutsche Grammophon, 1999); *Boulez: Répons,* directed by Robert Cahen (Colimason, INA, IRCAM, 1989), http://www .heureexquise. org/video.php?id=1188.

13. Stephen Healey, "Air Conditioning," paper presented at the Materials: Objects: Environments workshop, National Institute for Experimental Arts (NIEA), Sydney, May 19, 2011.

14. David Gissen, *Subnature: Architecture's Other Environments* (New York: Princeton Architectural Press, 2009), 79; "Reflux: From Environmental Flows to Environmental Objects," paper presented at the Materials: Objects: Environments workshop, NIEA, Sydney, May 19, 2011.

15. R&Sie, *Dusty Relief,* http://www.new-territories.com/roche2002bis.htm.

16. Neil A. Manson, "The Concept of Irreversibility: Its Use in the Sustainable Development and Precautionary Principle Literatures," *Electronic Journal of Sustainable Development* 1.1 (2007): 3–15, https://sustainability.water. ca.gov/documents/18/3407876 /The+concept+of+irreversibility+its+use+in+the+sustainable.pdf.

17. Fernand Braudel, *Civilization and Capitalism, 15th–18th Century,* trans. S. Reynolds, 3 vols. (Berkeley: University of California Press, 1982–84).

18. Aristotle, *Metaphysics,* 213, 217.

19. Marx, *Capital,* 1:620.

20. Burtynsky, *Manufactured Landscapes; Manufactured Landscapes,* directed by Jennifer Baichwal (Foundry Films, National Film Board of Canada, 2006).

21. Slavoj Žižek, *Enjoy Your Symptom! Jacques Lacan in Hollywood and Out* (New York: Routledge, 2001), 209.

22. ABCnews, "Oil From the BP Spill Found at Bottom of Gulf," September 12, 2010, http://abcnews.go.com/WN /oil-bp-spill-found-bottom-gulf/story?id=11618039.

23. Levi R. Bryant, *The Democracy of Objects* (Ann Arbor: Open Humanities Press, 2011), 208–27.

24. Mary Ann Hoberman, *A House Is a House for Me* (New York: Puffin Books, 2007), 27.

25. Hoberman, *House,* 34, 42–48.

26. Harman, *Tool-Being,* 68–80.

27. The phrase is Graham Harman's: *Guerrilla Metaphysics: Phenomenology and the Carpentry of Things* (Chicago: Open Court, 2005), 23, 85, 158, 161.

28. Joan Stambaugh, *Finitude of Being* (Albany: SUNY Press, 1992), 28, 53, 55.

29. An exemplary instance is Rocky Flats Nuclear Guardianship: http://www .rockyflatsnuclearguardianship.org/.

30. Thomas A. Sebeok, *Communication Measures to Bridge Ten Millennia* (Columbus, Ohio: Battelle Memorial Institute, Office of Nuclear Waste Isolation, 1984).

31. *Into Eternity,* directed Michael Madsen (Magic Hour Films and Atmo Media, 2010).

32. Susan Garfield, "'Atomic Priesthood' Is Not Nuclear Guardianship: A Critique of Thomas Sebeok's Vision of the Future," *Nuclear Guardianship Forum* 3 (1994): http://www.ratical.org/radiation/NGP/AtomPriesthd.txt.

33. Martin Heidegger, *Contributions to Philosophy: (From Enowning),* trans Parvis Emad and Kenneth Maly (Bloomington: Indiana University Press, 1999),13.

34. See Timothy Clark, "Toward a Deconstructive Environmental Criticism," *Oxford Literary Review* 30.1 (2008): 45–68.

35. Stambaugh, *Finitude of Being,* 93.

36. Derek Parfit, *Reasons and Persons* (Oxford: Oxford University Press, 1984), 355–57, 361.

37. Parfit, *Reasons and Persons,* 309–13.

38. Jacques Derrida, "Hostipitality," trans. Barry Stocker with Forbes Matlock, *Angelaki* 5.3 (December 2000): 3–18 (11).

39. Donna Haraway, *When Species Meet* (Minneapolis: University of Minnesota Press, 2007), 19, 27, 92, 301.

40. Rob Nixon, *Slow Violence and the Environmentalism of the Poor* (Cambridge, MA: Harvard University Press, 2011), 2.

41. Emmanuel Levinas, *Totality and Infinity: An Essay on Exteriority,* trans. Alphonso Lingis (Pittsburgh: Duquesne University Press, 1969), 160, 258; *Otherwise than Being: Or Beyond Essence,* trans. Alphonso Lingis (Pittsburgh: Duquesne University Press, 1998), 3.

42. Franz Kafka, "The Cares of a Family Man," in *Metamorphosis, In the Penal Colony, and Other Short Stories* (New York: Schocken Books, 1995), 160.

43. Kafka, "Cares," 160.

44. Kafka, "Cares," 160.

45. Chögyam Trungpa, *Glimpses of Abidharma* (Boston: Shambhala, 2001), 74; Heidegger, *Being and Time,* 171–78.

46. *The Matrix,* directed by the Wachowskis (Village Roadshow Pictures and Silver Pictures, 1999).

47. Radical Joy for Hard Times, "What Is an Earth Exchange?," http://www.rad icaljoyforhardtimes.org/index.php?option=com_content&view=article&id=79& Itemid=29. A slide show of the 2010 Global Earth Exchanges

can be found at http://www.radicaljoyforhardtimes.org /index.php?option=com_content&view= article&id=55&Itemid=5.

48. Morton, *Ecological Thought,* 38–50.

49. "Cosmic Origins: A Series of Six Lectures Exploring Our World and Ourselves," University of Arizona College of Science, http://cos.arizona.edu /cosmic/.

50. I refer to the movie *The Island President,* directed by Jon Shenk (Samuel Goldwyn Films, 2011), about Mohamed Nasheed, president of the Maldives, whose islands are being inundated by the effects of global warming.

29

My Father Pluto

ROBERT PHOENIX

Pluto was discovered on February 18, 1930, in Flagstaff, Arizona, just slightly above the 33rd degree parallel, where the bomb was first exploded at White Sands, where Phoenix lies just to the south, also at 33 degrees. It is parallel Memphis, where MLK was shot and the dream became a nightmare.

In astrological parlance, being discovered on February 18, 1930 would make it an Aquarian planet and, thus, its influence would extend beyond its mere mundane or common mythological meaning. Pluto would be forever tied to the death of culture and, since it was at 17 degrees Cancer, just 4 degrees off the USA Natal chart at 13, it will forever be linked to the land of Manifest Destiny.

Pluto, the planet of Scorpio, harbinger of life, death and rebirth, would be providing the kindling for the eventual immolation of the West.

But that's not really what I want to write about, although mine is an American story, as American as Pluto.

I lived with Pluto—he was my father.

My father was born five years after the discovery of that dark and icy orb. He was born in Shreveport, Louisiana. He was Steinbeck-worthy. His family was like that of many Dust Bowl immigrants who fled to the sunny state of California, where his mother would divorce his own alcoholic father and throw my father into a cyclonic downdraft of foster homes, rejection, and ultimately, neglect.

He grew up as a raging fury, fighting whenever anyone crossed eyes with him.

He left school early and enrolled in the Army.

His was given a dishonorable discharge after six months.

He drifted, fought some more, and then realized, in true Plutonian fashion, that he needed to re-enlist and complete the Plutonian/Phoenix cycle.

He became a model soldier, read the classics, learned some judo, and met my mother at an Army base dance.

My father had Mars and Jupiter in Scorpio, whose ruling planet is Pluto.

My first taste of Pluto was when I was around four and witnessed him knocking out my grandfather and my uncle in a drunken fistfight on a hot New Jersey night.

I was shuttled to a neighbor's house around midnight. We left New Jersey shortly thereafter.

The flight of the Phoenix across America took us to San Bruno, or Saint Bruno, the Catholic icon whose claim to fame was the eloquence he displayed in funeral rites and eulogies, which is a very Plutonic gift indeed, bestowing poetic blessings upon the newly dead.

Saint Bruno was canonized on February 17, 1623. That would be just one calendar day before Pluto was discovered on February 18.

San Bruno was the entry point to my journey into Hades. Say your prayers, oh sainted one—for the annihilation of my tender psyche was about to commence.

For the next ten years of my life, I would watch my father literally leave his body and become someone else, something else, and unleash the fury of hell. He knew it too, and there was nothing he could do about it. The neglect and rejection that he experienced as a child had been locked away in his Plutonian depths and could no longer be contained.

The relatively safe, if improper, alchemy of the home was the place where a damned transmutation could occur at a moment's notice. The domestic alembic became a Plutonic cauldron.

Astrologically his Mars in Scorpio was opposite his mother's Taurus Sun in a fixed alignment. For years, the shame that his mother felt about divorcing his father would be projected onto my father. Stubborn as a bull, only when she was about to receive Pluto's kiss on her deathbed would she openly confess love for my father. The raging fires of Scorpio would finally be doused by the vapors of her final breaths.

In my own astrological chart I have Pluto conjunct my Mid Heaven, I am a Scorpio rising and have Mars in my Eighth House. His Scorpio Mars was in my Twelfth House and, if you know anything about astrology, it is the most violent aspect of Mars in a chart: astrologers often call it "the placement of hidden enemies."

Over time, I learned to determine the tenor, timbre, and overall mood of the situation, testing the emotional atmosphere around me, sensing whether or not Hades would make a guest appearance. It helped me hone my instincts and psychic awareness in such a way that I could walk into a room and read whom was a threat or not. I could sense who had power and, later on psychic and intuitive levels. I could derive more subtle details. What was a bare survival skill at home turned into a useful tool in the world at large. Hypervigilance morphed into psychic awareness.

I began to learn how to live with both violence and intensity and extreme conditions because when my father was right, he was one of the best guys on the planet but, when he wasn't, he was sheer hell.

Astrologically, that's an aspect of Pluto as well, the light and dark, with no middle ground. The light switch was either on or off. The gray area was my own world, trying to make sense of the extremes.

Later in life, I realized that I was being put through some type of initiation during those formative years. I was experiencing things that no child should have to and yet I had come to realize by trial that I had a capacity to handle dark psychic material that would make most people blanch.

I could hold space for crisis, conspiracy, and catastrophe that set me apart from others who were raised under more "normal" circumstances.

However this too was distorted in some ways as my choices for partners would also have some Plutonian current running through their lives. I once dated a *bruja* who had been a morphine addict and lived an almost vampire-like existence in a dark hovel at the fringe of San Francisco's Mission District.

I was drawn to Persephone archetypes that would lead me back down into the underworld that had been my psychic stomping grounds, my personal familiar hell. I would spend plenty of lost time in these places, exorcizing his demons and now mine.

What I learned was that the devils that rose up from my father's Plutonian depths were in some ways my teachers, the guides of hidden and desperate spaces, unspoken epiphanies, and quarters of the dark side. They were once and future gatekeepers of the shadow realm. Their shape-changing presence would continue on a personal and professional level as a theme in my life.

My father was an unconscious shaman of sorts, leading me into a lifelong exploration of the underworld and psychic depth.

In his honor, I want to share a positively Plutonian experience before anyone who reads this gets the impression that I lived with (Scorpio) Charlie Manson's doppelganger.

One day, I was at school and I got into a hassle with a kid in line and as a result, he wanted to, "call me out" as we used to say back in the day. I agreed and we were going to fight after school.

Well, my heart wasn't in it and I deftly avoided him. But he went to my house and asked my father if I was there. My father asked him why. He told him that we were supposed to meet after school and fight.

When I got home, my father asked me about him and if it was true. He also asked me if I was scared. I told him, "Yes." He then told me to tell him that I would fight him in two weeks.

The next day at school, I did. I told him.

For the following two weeks, my father trained me in the darkness of our suburban garage. He strung up a heavy duffle filed with clothes, which became our training bag. He put up a speed bag too and taught me how to hit it. We skipped rope, practiced combinations, and lifted weights. Here was his Mars in Scorpio as an ally, not the tempestuous and hidden enemy stalking me in the hidden spaces of my twelfth house. In the dark square, upon slab concrete, Pluto assumed form as my own, personal trainer.

Two weeks later, I walked over to my adversary's house on a cul-de-sac, a dead end, a Plutonian semi-circle of sorts.

I knocked on his door and his mother answered.

"Are you here to fight David?"

Clearly she was in on the deal. "Yes, I am," I said, being in on the deal too.

David strode out of the house, barrel-chested with bravado, and met me in the center of Pluto's cul-de-sac.

He wasn't ready for the ninth planet. I put on a boxing clinic, moving and jabbing, sticking and feinting. It wasn't a fair fight. I dodged his clumsy haymakers as if they were being thrown in slow motion. I didn't so much beat him up as humiliate him. Out of frustration he started to kick me, and his mother, ever the overseer, the Moon, watching it all from her window, came out and stopped the fight.

She said, "Your father taught you how to fight, didn't he?"

I didn't know what she was getting at, but my Pluto must have known the entire act because I said, "Yes, but he also taught me how to make friend afterwards." I stuck out my hand to shake David's. He reciprocated and the two went back inside.

A month later, they moved.

So you see, Pluto is complex, multi-layered, and relentless from an astrological perspective. It's not for novices or first-timers. It's for those of us who are likely not going to change in any other way, who won't budge unless we are confronted by the intensity, violence, and potential for rebirth seminal in the moments of utter darkness.

I would not be who I am today, for better or worse, without Pluto, Pluto in my chart and Pluto in the guise of my Dad, and I love him for it. Father might have known best, but perhaps Pluto knew better.

30

Pluto is the Reason We Have a Chance

An Interview*

Ellias Lonsdale

Pluto is literally dropped into the deepest layers of consciousness into which we are capable of dropping, that we can conceive of, that we can imagine, to which we can penetrate, that we can actually probe. For most people most of the time, Pluto is unconscious, yet compulsive. It comes out of nowhere; they don't know what it is, they don't work with it very well. But if you happen to be a depth-charged soul, Pluto is the greatest of allies because it gives you coordinates to the underworld, to inner worlds, to other worlds. It provides a journey to take. It gives you core resonance, a kind of radical evolutionary path, which is different for every being, every person.

Pluto has a tendency to be overwhelming and yet perfect. Its perfection is something you discover late in the game, but its overwhelmingness is something you're with all the time.

I would say that it is because of Pluto that we have a chance. Every other planet, relatively speaking, even Neptune, is superficial, is surfacy, is plugged into programs and conditionings, habits and syndromes of memory. Pluto is able to pierce the veil, every veil, including the veil of collective consciousness and the way we usually think about things, the way we usually operate, which is asleep, tuned-out, absent, missing. Pluto is what brings us here, brings us through; it is overwhelmingly the one planet you can depend on. You can depend that Pluto will show up. It will do stuff. It won't be among the missing, it won't be vague, it won't be nebulous.

My sense of it is that if people were fully aware of what Pluto is doing, they would treat it like the devil, they would treat it as something to exorcize

---

*. Interviews were conducted by asking an author to talk about Pluto and taping/transcribing his response. Though edited for basic readability, they are verbal, not literary, documents. They should be heard aloud as speech.

or get rid of, to avoid or run from, because it defies logic, it contradicts our way of life. Pluto forces us to surrender, to let go, to enter a different existence, to learn about things, remember things, realize things, accept things that we typically prefer to avoid.

Pluto is more experiential than even your most powerful and poignant experiences. It tells you that have to be there, you have to earn it, you have to earn your existence. You have to be absolutely in it and through it. You cannot talk it, you cannot fake it, and you certainly cannot capitalize on it. Because it knows all that, because it is the *fact* of that, Pluto is the ultimate equalizer.

Pluto is the best place to begin the infinite journey, for it will kill off any ego-complexes you happen to still observe. It will keep you honest and make you learn the ways of the Earth. When the Earth is taken as a cosmos, as a total system, it becomes Pluto's special province, special concern, special interest, because living in the Earth is reincarnational, is physical, is the fully-embodied reality, potentially, that Pluto is trying to facilitate and work us into and through. In that sense, Earth is Pluto, and Pluto is Earth. Earth is the depth of Pluto, and Pluto is the mystery of Earth.

Where astrology and astronomy meet, Pluto is a kind of an abyss, a void-space, almost like a black-hole kind of place that's at the center of a vast other realm. It's the king and the focal point of that realm but utterly elusive in everyday systems and circumstances. You'd really have to keep digging and digging, probing and probing, penetrating and excavating to get at it.

31

Pluto: Planet of Wealth

An Interview[*]

ROB BREZSNY

When I talk about Pluto, I'm not picturing a god, I'm not picturing a man—I'm picturing a realm. The god Pluto can stand in and be symbolic for a place. The place is known by many names: the Underworld, the astral plane, the fourth dimension. I call it the other *real* world—that's one name I have for it. It's a universe ruled as much by Persephone as Pluto. In fact, for me, Persephone precedes Pluto. The Persephone-Pluto dyad is archetypally symbolic of a whole sphere of topography and being. So when I ask myself, "What is the influence of, or what is the meaning, astrologically, of the planet Pluto?" I automatically add, "What is the meaning of the planet Pluto/Persephone?"

Before there was a masculine god Pluto in ancient mythology, before he governed a subterranean realm, Persephone was its sole ruler. Only a later patriarchal gloss enthroned Pluto there, but his deposing of Persephone required a violent act of rape and kidnapping, which is how that she got taken below. The patriarchal version of the myth is indicative of the long-standing subversive nature of Pluto: Pluto wouldn't be the Underworld or Hades without an interloper dragging Persephone there. But the Pluto/Persephone archetype transcends any one version of the Pluto/Persephone myth. In other words I am talking about a deep, hidden force that manifests throughout the universe on different worlds in different times and is always seminal. You can't have Pluto without Persephone, but you also can't have Persephone unless Pluto awaits her.

---

[*] Interviews were conducted by asking an author to talk about Pluto and taping/transcribing his response. Though edited for basic readability, they are verbal, not literary, documents. They should be heard aloud as speech. (Paul Foster Case's pattern on the Trestleboard, referred to by name in Brezsny's riff, was added later.)

For me, living in the Plutonian realm is a key part of being human. You need to dwell at two levels: the surface, which is the domain of the ego, of transient wealth and shiny material things, and the Underworld or what the Qabala calls Foundation or Yesod. I don't mean to suggest that ego experiences are lesser or should be devalued as such. It's just that they have become grossly overexaggerated and overemphasized in our culture. As an advocate for the realm of the soul, I am always overcompensating on its behalf. A healthy human being must be able to go back and forth between the two realms and acquire both sorts of experiences and knowledge. As is, no matter how ego-based our culture, we have to spend at least seven or eight hours every day in the Underworld. There we meet Persephone, in sleep and in dream, as we go back to the Foundation, to our source, and dwell there for a spell.

I first began noticing and paying attention to Pluto around age thirteen, though not by name, as I started remembering and writing down my dreams. I recognized my need for a vital connection with the Underworld, and four years later I became a feminist and goddess-worshipper. I was seeking modes of intelligence that weren't being taught at school.

Unfortunately, our modern culture is more and more ignoring, denying, repressing, banishing Pluto. This is particularly obvious in the rise of machine consciousness: we are being infected by a mechanistic way of being, thinking, and feeling. In our overly logical and rational minds, technology has become the sole viable or authorized mode of intelligence. At the same time, the realm of the soul—the way in which our organism functions as soul—is increasingly demeaned and diminished. The loss is immeasurable, for the soul is essential to our survival.

Don't get me wrong: the scientific rubric is of great value; it's an important mode, powerful and beautiful in its way. But a pathology develops when it's the only mode recognized as important or valid, when nothing else is really real and nothing else needs to be nourished, nothing else leaves its imprint on our existence. When we try to live as machines or the familiars of machines, even our emotional lives become mechanistic.

Symbolically this situation is illustrated by the 2006 demotion of Pluto from planetary status. Scientists no doubt had sound, rational reasons to exile Pluto from the traditional Solar System and transfer its realm to the Kuiper Belt with the other dwarf worlds, but they were also under the influence of deeply unconscious forces too. The expulsion of Pluto marked a symbolic turning point in the triumphalism and triumph of scientism, the ascendency of a mode of thinking that values only what's visible, measurable, and categorizable. But Pluto is more than the rocky planetoid representing it: Pluto is an essential phase of human consciousness.

It is no accident that ostracism comes at a time when the Plutonian realm itself is being devalued and rendered inessential. The overall downgrading of Pluto is a milestone in the modern attempt to depreciate the soul's mode of awareness and make it subsidiary to the deductive mind. To banish Pluto is to deny that living in the soul has any value to us.

Though Pluto of course existed before its discovery by astronomers in 1930, 1930 was the year in which its archetype came into our awareness. I wonder if there is something that the first appearance of Pluto in the conscious realm corresponded to in the Underworld from which it came. A culture bent on denying the Underworld's existence inherits the Underworld's chaos and pandemonium instead. In this case the forces leading World War II gathered as a new planet entered the zodiac.

But Pluto appeared when we needed it. Many Plutonian phenomena and influences in astrological charts can be traced to the ways in which we require Pluto but aren't getting it. The harder that atheistic science practitioners try to scour all evidence of the Underworld, the more Pluto's effects are going to manifest in bizarre and violent forms. They appear, like Pluto, because they are there.

We have plenty of examples of negative Pluto. Religious fundamentalism, climate-change denial, accumulations of obscene amounts of wealth by the already wealthy are among the pathological emanations of Pluto, the result of its realm being suppressed. Financial instruments that are developed solely to create monetary value at the expense of labor and creativity elevate individuals who haven't earned their assets and have no idea of what to do with them except to create more. Bernie Madoff was a classic example of Plutonian madness: he already had enough money for three lifetimes when he committed the crimes that got him imprisoned in a Hades-like cell.

But Pluto is neither the transgression nor its jailer; it is the sash that oscillates back and forth between the insanity of ego-derived and factory-generated, media-hyped wealth and the depth of soul wealth. Plutonian treasures are often the antitheses of economic value and lead to the alchemical transformation of money and assets into a different, more organic currency that the soul can actually assimilate and use.

July 2015, when NASA's *New Horizons* satellite will arrive at the outermost original planet, marks the final transit of the big Pluto-Uranus square that has been unwinding since 2009. By 2015 it will have taken place six or seven times. Some astrologers attribute the financial collapse of 2008 to the direct influence of that square. That is also why they worry that there's still more turmoil to come: the tension between Pluto and Uranus hasn't completely climaxed.

The trickster is a crucial element of our historic engagement with the Pluto archetype because it goes back and forth between conscious and unconscious realms with ease. Practitioners of scientism and the sciences don't like the trickster because it messes with boundaries, it dissolves definitions, it makes it hard to tell what's real and what's not. But the trickster is going to show up whether atheistic science practitioners want him or not, and increasingly he's creating a jumble of ideas and phenomena that leave scientists completely confused and enraged. Lately they have been as outraged about homeopathy, probiotics, and UFOs as their foes have been about claims of melting glaciers, the biology of gender, and evolution. Pluto's trickster universe is becoming one big maddening, irrational soup for mainstream science. Yet to the degree that they keep banishing Pluto, the trickster is going to get bigger and bigger and bigger, in bizarre and cartoony ways too.

The modern world doesn't just try to convince us that the soul doesn't exist. Convincing us is the single purpose of modernity. We want to conquer Pluto's shadow world and get it away from trickster deeds and signs. We want to pretend that Pluto doesn't exist or, if it does, that it is not important enough to be granted a kingdom.

But it's really the soul that is being banished, that's on a trail of tears headed out to some backland, some faraway place—it doesn't actually matter where as long as it won't bother us anymore, as long as it won't insert its crazy wisdom into our conversations, as long as it won't get in the way of our hoarding false wealth and putting the universe into privately-owned boxes and cubicles.

The modern reaffirmation of the classical myth of Pluto merely ratifies the canard or slander about Pluto—that we can only go down to its territory by rape and abduction, that only through an act of violence and violation can we get our experience of the Underworld, of the soul's realm. The soul, which is said not to exist, is allowed back only by trespass and transgression. That is Pluto's new, unspoken myth.

But of course if we deny, banish, and repress all the soul's good stuff, the only way it can reach us is through rape and abduction. As its cycle of violence and prurience increases, Pluto imposes a daily punishment, which is not actually a punishment but a necessity, a gift in disguise. The soul will not allow us to forget it, even if it has to express itself by dark acts of requital. Pluto's revenge is the revenge of a world that's being denied. People are taken to their ravaged and forgotten souls by acts of mutilation and pornography. So we end up relating to Pluto through violent means rather than by daily familiarity and engagement with the archetype itself or by becoming aware of the soul's radiance in quiet, inconspicuous, mundane ways.

The soul is the root of this world; that's what shamans in every culture profess. That's where they go to find healing that will repair the body in this world. The Qabalists tell us that Yesod is the origin and foundation of this plane of creation. Paul Foster Case's pattern on the Trestleboard is an homage and prayer to Pluto:

0. All the Power that ever was or will be is here now.
1. I am a center of expression for the Primal Will-to-Good which eternally creates and sustains the Universe.
2. Through me its unfailing Wisdom takes form in thought and word.
3. Filled with Understanding of its perfect law, I am guided, moment by moment, along the path of liberation.
4. From the exhaustless riches of its Limitless Substance, I draw all things needful, both spiritual and material.
5. I recognize the manifestation of the Undeviating Justice in all the circumstances of my life.
6. In all things, great and small, I see the Beauty of the Divine Expression.
7. Living from that Will, supported by its unfailing Wisdom and Understanding, mine is the Victorious Life.
8. I look forward with confidence to the perfect realization of the Eternal Splendor of the Limitless Light.

9. In thought and word and deed, I rest my life, from day to day, upon the sure Foundation of Eternal Being.

10. The Kingdom of Spirit is embodied in my flesh.

Pluto, the soul, Yesod abide; they're there in every single moment—as treasure, as possibility. And to the degree to which we ignore them and don't look for them and only value things in the material world, to that degree we build up the necessity for Plutonian shocks; for rapes, abductions, and seizures; for events that force us to go down and look at what's beneath, what's in the core of our being, what's below us at the roots of this world.

When Thomas Moore and James Hillman talked about soul work, what they meant was little acts of unions with the underneath part of the world. It is always showing through if we'll just look for it. Not spectacular actions, ecstatic dances at a Burning Man event, though that's fine. Or some crazy Tantric fuck, which is fine too. I have nothing against those Pluto manifestations. Or God forbid, disease and death—that's another sure way of getting to Pluto.

You might recall that the designer L'Wren Scott, Mick Jagger's girlfriend, committed suicide in early 2014. My wife Roe and I looked at her chart and saw that Pluto was transiting over her Sun, right on her Sun in fact. The consensus seems to be that she committed suicide because, in large part, she was a failure financially, deep in debt.

I don't want to speculate at her expense, but this may have been a classic Plutonian suicide—a rush, a sudden appearance, of the soul that could not be tolerated because it was too strange and horrific and empty. The realm in which this woman was admired and achieved an identity as a designer of pretty things that cost a good deal of money was something that her ego could no longer sustain. It didn't match what was really happening in her soul.

The emotional breakdown was an invitation for her, and again I speculate, to go down below and find another way to value herself; she didn't have to be only a designer of pretty clothes for the wealthy; she could become something more real.

But she couldn't handle the invitation. She didn't want to deal with the loss of this other kind of wealth, the ego's wealth. She refused Pluto's invitation to experience the different wealth of the Underworld.

When people only derive riches through material objects, power, money, what the ego values, they are sending an irresistible invitation for Pluto to bring them down.

We can cultivate a more ordinary Plutonian ritual and mantra too. For instance, as I'm walking the path today that I always walk, I notice that one of my favorite rocks is gone. What happened to it? There's some mystery there. At the moment I recognize the rock's movement, I'm taken into the realm of soul. I've been mystified by a manifestation from the Underworld. On my mundane walk through the natural world, daily life has been deepened and the realm of the soul has inserted itself, even as a lacuna. That's how I prefer to relate with Pluto.

Even if you're a good boy or girl and you follow the soul's daily plan, you might still be dragged down by rape and abduction. It's an active universe, after all. Everything is always a possibility. We can't be so arrogant to think that we know what Pluto's going to do with us or how it's going to do it, how we're going to get into its realm.

But daily Plutonian observation is an opportunity to visit the Underworld and the soul on a regular basis. It is one of uniquely meaningful aspects of being embodied in this form, alive. If we ignore it or neglect it and overvalue what the ego does as all-important, what capitalism produces as the lone authentic wealth, what materialism does is as the whole show, and start to locate our values in machine glitter and economic affluence, we forfeit the subtler riches available in the Plutonian depths.

Pluto is the actual source of treasure, not from the ego's point of view but from the soul's perspective. For the soul, Pluto is the realm of riches.

Pluto can seem like the worst thing that ever happened to you when it is actually the best thing that could ever happen to you. But you have to recognize it as what it is rather than by the robes and kitsch death symbols in which it appears. You have to meet it in the actual realm of the soul.

About the Contributors

College of the Atlantic professor **Rich Borden** offered the Pluto question-naire as an exercise for his class "Ecology and Experience." Several students responded, some by name, some anonymously. A few of their entries are curated here.

Rob Brezsny has been a rock composer, musician, and performance artist (Tao Chemical and World Entertainment War), an astrologer (freewillas-trology.com), and a literary writer (a novel, *Televisionary Oracle;* a poetry col-lection, *Images Are Dangerous;* and an oracle, *Pronoia is the Antidote to Paranoia: How the Whole World is Conspiring to Shower You with Blessings*). He is also the author of the signature 1970s piece "Qabalistic Sex*Magick for Shortstops and Second Basemen," which was published in *Baseball I Gave You All the Best Years of My Life* (Oakland, California: North Atlantic Books, 1978). He lives in the San Francisco Bay Area. He can be reached via freewillastrology.com.

Fritz Brunhübner, a life-long professional astrologer, was born on July 3, 1894 in Nurnberg, Germany, and died on March 9, 1965. Author of the book *Pluto,* he wrote, "We have to consider the connections between Pluto and Uranus more thoroughly as these are very strong in their action, which started in 1930 and brought a continuous shaking and trembling of the Earth and its inhabitants."

Maggie Dietz is the author of the award-winning book of poems *Perennial Fall* (University of Chicago Press) and the former director of the Favor-ite Poem Project, Robert Pinsky's special undertaking during his tenure as U.S. Poet Laureate. Her other awards include the Grolier Poetry Prize, the George Bennett Fellowship at Phillips Exeter Academy, as well as fellowships from the Fine Arts Work Center in Provincetown and the New Hampshire State Council on the Arts. Her work has appeared widely in journals such as *Poetry; Ploughshares; Agni; The Threepenny Review;* and *Salmagundi.* "Pluto" appears in her forthcoming collection, *That Kind of Happy.* She teaches at the University of Massachusetts, Lowell.

Thomas Frick is a writer and editor living in Los Angeles. His alchemical Luddite novel *The Iron Boys* was published in November 2011, on the two hundredth anniversary of the first Luddite attack. *The Sacred Theory of the Earth*, an anthology he compiled of texts and art concerning metaphorical and esoteric dimensions of landscape, was published by North Atlantic Books in 1986.

A native of New York City (1944), **Richard Grossinger** attended Amherst College and the University of Michigan, receiving a BA in English (1966) and a PhD in anthropology (1975). He wrote his doctoral thesis on his fieldwork with fishermen in Eastern Maine, after which he taught for two years at the University of Maine at Portland-Gorham and five years at Goddard College in Vermont. He is the author of numerous books, including *New Moon; Planet Medicine; The Night Sky: Soul and Cosmos; Embryogenesis: Species Gender and Identity; Embryos, Galaxies, and Sentient Beings: How the Universe Makes Life; On the Integration of Nature: Post-9/11 Biopolitical Notes; The Bardo of Waking Life; 2013: Raising the Earth to the Next Vibration;* and *Dark Pool of Light: Reality and Consciousness.* With his wife, Lindy Hough, he is the cofounder and publisher of North Atlantic Books as well as its forerunner, the journal *Io.*

Ross Hamilton is a writer specializing in archaeological antiquities, their form and function, and their astronomies: what we may learn from them that might benefit us today. He has published twice on the subject, both times through North Atlantic Books: *The Mystery of the Serpent Mound: In Search of the Alphabet of the Gods* and *Star Mounds: Legacy of a Native American Mystery.* He volunteers and works at Ohio's Serpent Mound Park during the warm weather.

Stephan David Hewitt is a spiritual counselor and astrologer, singer and composer *(Full of Life Now: Love Songs of Walt Whitman),* as well as the publisher and editor of *Hand to Hand,* a community-based endeavor that supports independently published works. He writes a "New and Full Moon" astrology blog every month at www.stephandavidhewitt.com. He lives in Santa Monica, California, and the Big Island of Hawai'i.

James Hillman (April 12, 1926–October 27, 2011) was an American psychotherapist (and critic of psychotherapy) who practiced in, while enlarging, the Jungian tradition. He was trained at, and later guided studies for, the C. G. Jung Institute in Zurich, Switzerland. His books include *A Terrible Love of War; The Soul's Code: On Character and Calling; Healing Fiction; Re-Visioning Psychology; Pan and the Nightmare; Suicide and the Soul; Anima: An Anatomy of a Personified Notion, Senex and Puer;* and *The Dream and the Underworld* (from which the selection in this anthology is plucked).

For over a generation, **Richard C. Hoagland**—participant in the Apollo Lunar Landing Program, former science advisor to Walter Cronkite during the Apollo Program, and former NASA consultant to the Goddard Spaceflight Center following the end of the Apollo program—has scientifically pursued the growing possibility of "a former, extraordinarily advanced, ancient ET civilization previously inhabiting . . . and reshaping . . . the original solar system."

Hoagland's two major works detailing this decades-long, multidisciplinary research effort (*The Monuments of Mars: A City on the Edge of Forever* and, with Mike Bara, *Dark Mission: the Secret History of NASA*) have both become international best sellers, published in a variety of updated editions and languages. In 1996, Hoagland founded The Enterprise Mission—a not-for-profit, public policy research organization charged with continuing this groundbreaking extraterrestrial investigation, including its most profound, wide-ranging implications.

A former NASA astronaut, **Jeffrey A. Hoffman** is currently professor of the Practice of Aerospace Engineering in the Department of Aeronautics and Astronautics at the Massachusetts Institute of Technology in Boston. Hoffman made five flights on the space shuttle, including the first mission to repair the Hubble Space Telescope in 1993 when the orbiting observatory's flawed optical system was corrected.

Shelli Jankowski-Smith's poems and essays have appeared in numerous publications including *Agni; the Boston Globe; CrossCurrents; Elephant Journal; Harvard Review;* and *Salamander.* She holds an MA in Creative Writing from Boston University. Shelli works as a Reiki Master Teacher, healer, and meditation guide as the owner of Sunflower Reiki and Wellness in Swampscott, Massachusetts.

Robert Kelly teaches in the Written Arts Program at Bard College. His most recent publications are the long poem *Uncertainties*; the novel *The Book from the Sky;* the collection of short fiction *The Logic of the World;* a collection of five plays, *Oedipus after Colonus and other plays;* and *Winter Music,* texts to the photo work of Susan Quasha. His collaboration with the painter Nathlie Provosty, *The Color Mill,* and his *Collected Essays* (edited by Pierre Joris and Peter Cockelbergh) were published in Fall 2014.

Jonathan Lethem is a novelist, essayist, and short-story writer in a genre that mixes elements of detective stories and science fiction. His works include *Gun, with Occasional Music; The Fortress of Solitude; Amnesia Moon; As She Crawled Across the Table; Motherless Brooklyn; and Dissident Gardens.* A native of New York City, he is currently Roy Edward Disney Professor of Creative Writing at Pomona University.

A native of New York City and graduate of SUNY at Binghamton, **Ellias Lonsdale** is a psychic and astrologer based on the big island of Hawai'i. A student of Dane Rudhyar, Marc Edmund Jones, and the Steinerite astrosophers, he taught at a Waldorf school in Auburn, California, in the 1980s and was the founder of an Atlantean mystery school in Santa Cruz, California, in the 1990s. His books include *Inside Planets; Inside Degrees; Inside Star Vision; The Book of Theanna (In the Lands that Follow Death); The Christ Letters;* and *Cosmic Weather Report: Notes from the Edge of the Universe* (with Mark Borax). He can be reached for consultation, charts, and readings at www.stargenesis.com or by leaving a voicemail at (831) 425-3134.

Steve Luttrell was born and continues to live in Portland, Maine. He is the founding editor of *The Cafe Review,* a quarterly journal of poetry and visual art. He has done five collections of poetry and several chapbooks. His newest title is *Plumb Line,* released by North Atlantic Books in 2015.

J. F. Martel is a writer and filmmaker based in Ottawa, Canada. He is a contributor to the web magazine *Reality Sandwich* and the author of *Reclaiming Art in the Age of Artifice: A Treatise, Critique, and Call to Action,* published by Evolver Editions/North Atlantic Books in 2015.

Timothy Morton is Rita Shea Guffey Chair in English at Rice University in Houston, Texas. He is the author of many books, including *The Ecological Thought; Ecology without Nature;* and *Hyperobjects: Philosophy and Ecology after the End of the World,* from which his selection in this anthology is excerpted.

Charley B. Murphy is the author of the novels *Cute Eats Cute; End of Men;* and the upcoming *Bardo Zsa Zsa.* He is a pop surrealist painter and an occasional cartoonist. His website is www.cbmurphy.net.

Robert Phoenix is an Austin-based astrologer, radio host, alternative researcher, writer, father, and baseball coach (not necessarily in that order). He's been to Pluto between lifetimes and it looks surprisingly like Kurt Vonnegut's Tralfamadore sans Wildcat Montana. He owes much of his professional life to his Plutonian initiator a.k.a. his incarnational father, to which he dedicates this submission/entry into this fine anthology.

Dinesh Raghavendra is a twenty-five-year-old writer from Shimoga, India. He is a contributing reviews editor for the *Former People* journal (http://formerpeople.wordpress.com/) and a music reviewer for *Metalbase* (http://metalbase.in/category/reviews/). He works for a software company for a living and occasionally writes fiction, poetry, and essays. You can reach out to him by sending an email to dineshraghavendra@gmail.com.

Lisa Rappoport is a letterpress printer and book artist, producing limited edition artists' books and poetry broadsides under the imprint Littoral Press. Her work is widely collected and has been included in such surveys as *500 Handmade Books, Vols. 1 & 2* (Lark Books) and *1000 Artists' Books* (Quarry Books). She has two smallish books of poetry: *Words Fail {Me},* published by the San Francisco Center for the Book (2014), and *Aftermaths/Figments,* a chapbook from Etherdome (2009). Her website is littoralpress.com.

Gary Rosenthal is a poet and transpersonal psychotherapist who originally studied psychology at the Jung Institute-Zurich. His contribution here is culled from a forthcoming book that ranges in the interface of myth, psyche, and culture—*The Death of Narcissus: & Other Heresies for an Age of Narcissism.* Other published books include a collection of ecstatic love poems, *The*

You That is Everywhere, a poetry chapbook, *The Museum of the Lord of Shame,* plus another poetry collection, *An Amateur's Guide to the Invisible World.* He lives in the San Francisco Bay area where he also leads retreats, whacks golf balls, and consults with clients (via Skype) throughout the world. His website is www.garyrosenthal.net.

Robert Sardello is co-director of the School of Spiritual Psychology. He is also a faculty member of the Dallas Institute of Humanities and Culture, Texas; the Chalice of Repose Project, Missoula, Montana; and a former chairman of the Department of Psychology, University of Dallas. As a practicing psychotherapist for over twenty years, he has worked in Jungian and Archetypal Psychology. His books include *Silence: The Mystery of Wholeness; Love and the Soul: Creating a Future for Earth; Steps on the Stone Path: Working with Crystals and Minerals as A Spiritual Practice; Facing the World with Soul; Love and the World; Freeing the Soul from Fear;* and most recently *The Power of Soul: Living the Twelve Virtues.*

Nathan Schwartz-Salant is a psychoanalyst and author. Among his books are *The Mystery of Human Relationship* and *The Black Nightgown.* His major interest is the application of alchemical ideas to the here and now of clinical practice.

John D. Shershin is a designer/craftsman, woodworker, carver, and sculptor, specializing in custom building restoration and contemporary design and fabrication. He has been an amateur astronomer since his teenage years, viewing the night sky with his backyard telescope, and has been a long time student of astrology, Jungian psychology, and metaphysics. He studied and taught astrology at the Theosophical Society in Boston, Massachusetts, and is currently involved in exploring and invoking the imaginal realm in words and craft. He splits his time between the Boston area and a community group in western Massachusetts. He can be reached at shershinjohn@yahoo.com.

Jim Tibbetts has an MA in theology, STL (Licensiate of Sacred Theology) in Marian Studies (matters involving the Virgin Mary), and an MBA. He is also a mime, gives talks, leads retreats, and has written more than fifteen books, half of which are on plant-based nutrition, half on spirituality. His website is www.jimtibbetts.com.

Dana Wilde of Troy, Maine holds a doctorate from Binghamton University, where his dissertation covered intersections between science and the humanities. His writings have appeared in a wide variety of academic and popular publications, including *Studies in the Humanities; Mystics Quarterly; The Quest; Alexandria: Journal of the Western Cosmological Traditions; The Magazine of Fantasy & Science Fiction; Exquisite Corpse;* and many others. His most recent book is *Nebulae: A Backyard Cosmography.*

Philip Wohlstetter is a Seattle writer born in New York City who is old enough to remember when Pluto was a planet. He starred as Hamlet, yes *that* Hamlet, in a Columbia University production while an undergraduate there during the 1960s, played guitar on the streets of Paris after graduation, and was incarcerated and almost executed while traveling in Chile in reputed left-wing company during the junta overthrow of Salvador Allende. He is one of the authors of the collective 1983 detective novel *Invisible Seattle* (the "author" himself who has also dabbled in software, airplane-manufacture, and "grunge" music, continues to haunt the shores of Puget Sound in the Pacific Northwest). He is also the author of the virtual hypertext *Valparaiso,* which he has been composing in one form or another for forty years.